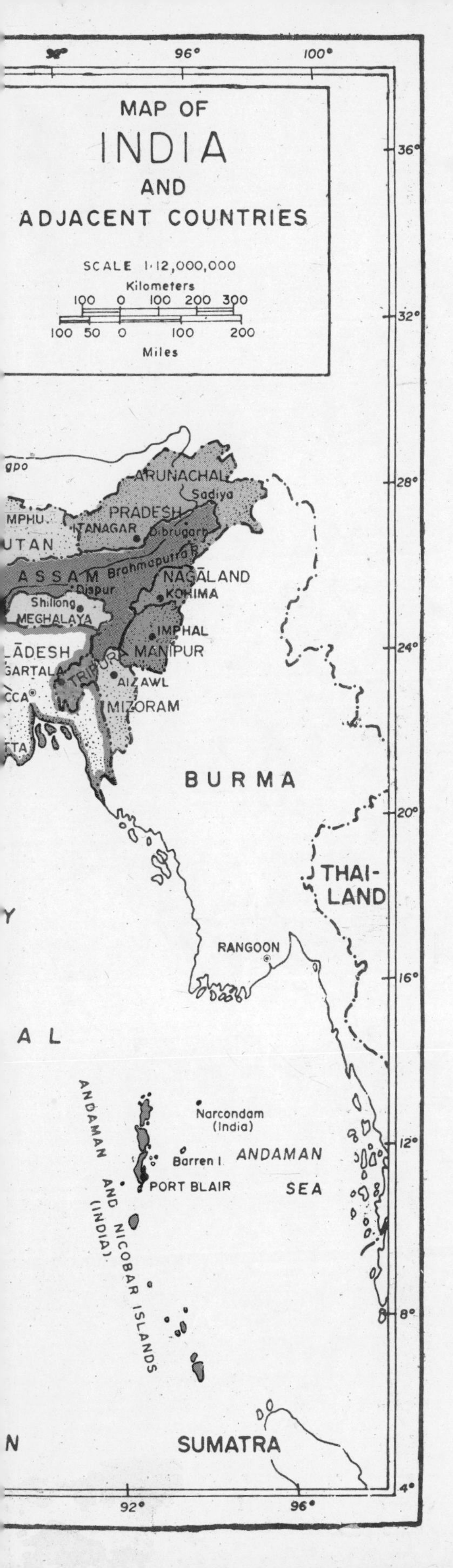

Based upon Survey of India map with the permission of the Surveyor General of India. The territorial waters of India extend into the sea to a distance of twelve nautical miles measured from the appropriate base line. The boundary of Meghalaya shown on this map is as interpreted from the North-Eastern Areas (Reorganisation) Act 1971, but has yet to be verified.

LIST OF TERMS USED FOR GEOGRAPHICAL AREAS OR TYPES OF TYPICAL FOREST

In the following pages certain Indian names have been used in referring to the habitats of birds. They are:

Bhabar. A type of high, dense sal forest containing also Sissoo *(Dalbergia)* and Semal *(Bombax),* found on alluvial and loamy soil with high rainfall (over 200 mm annually) extending from the northern edge of the terai to *c*. 600 m in the Himalayan foothills, characteristically in U.P. (Kumaon), but also east through Nepal to Assam.

Terai. The undulating, alluvial, often marshy strip of land stretching along the southern edge of the bhabar, between it and the north Indian plains, composed of grassland (sabai grass) and cultivated tracts, interspersed with dense tracts of forest. Found in U.P., Nepal, and east to northern West Bengal.

Duars. The easternmost parts of the terai, found in northern West Bengal (Jalpaiguri), Bhutan, and Assam in the adjacent areas to Bhutan.

Duns. The broad valleys within the outer ranges of the Himalayas, e.g. Dehra Dun (lying between the Siwalik Hills and the Himalayas proper, and between the Jumna and Ganga rivers), and the Patli Dun drained by Ramganga river. There are many other smaller and lesser known duns.

Sholas. Patches of montane wet temperate forest, of dense evergreen type usually in sheltered nallahs or hill-stream valleys amongst rolling grassland hills in South India (Nilgiri, Anamalai, and Palni hills particularly), and Sri Lanka, from about 1500 m up.

A PICTORIAL GUIDE to the BIRDS of the Indian Subcontinent

A PICTORIAL GUIDE to the BIRDS of the Indian Subcontinent

By SÁLIM ALI & S. DILLON RIPLEY
With 106 plates depicting all the birds
by JOHN HENRY DICK

BOMBAY NATURAL HISTORY SOCIETY

OXFORD UNIVERSITY PRESS
BOMBAY DELHI CALCUTTA MADRAS

Oxford University Press, Walton Street, Oxford OX2 6DP

Oxford, New York
Athens, Auckland, Bangkok, Bombay,
Calcutta, Cape Town, Dar-es-Salaam, Delhi,
Florence, Hong Kong, Istanbul, Karachi,
Kuala Lumpur, Madras, Madrid, Melbourne,
Mexico City, Nairobi, Paris, Singapore,
Taipei, Tokyo, Toronto,
and associated companies in
Berlin, Ibadan

First published 1983
Second impression 1989
Second Edition, updated 1995

ISBN 0 19 563732 1

PRINTED BY BRO. LEO AT ST. FRANCIS INDUSTRIAL TRAINING INSTITUTE, MOUNT POINSUR, BORIVLI (W), BOMBAY 400 103, PUBLISHED BY THE BOMBAY NATURAL HISTORY SOCIETY AND CO-PUBLISHED BY NEIL O'BRIEN, OXFORD UNIVERSITY PRESS, OXFORD HOUSE, APOLLO BUNDER, BOMBAY 400 039.

CONTENTS

INTRODUCTION

Specific identification is the ABC of meaningful bird watching as much as of scientific field research. Howsoever significant a field observation, its importance is lost unless the concerned species is correctly identified. For the untutored beginner good illustrations of birds, preferably in colour, are fundamental. The truth of this was clearly demonstrated by Hugh Whistler's pioneering *Popular Handbook of Indian Birds* first published in 1928, in creating and developing an interest in birds and birdwatching among the Indian public where it was practically non-existent before. Although the Popular Handbook contained only a few illustrations of the birds described in the text, and fewer still in colour, it triggered an immediate spurt of interest in birds and birdwatching, necessitating a second edition of the book in 1935 followed soon by a third edition and a fourth.

In 1941 Bombay Natural History Society (BNHS) first published *The Book of Indian Birds* by Sálim Ali describing 181 species of the commoner birds, all of which were shown in colour. The popularity of The Book, largely due to this feature, has necessitated further editions every few years, each edition enlarged progressively by the inclusion of a few more species, till the latest, eleventh, published in 1979, contains accounts and colour illustrations of 296 species. However, this represents merely a small fraction of our total avifauna, and it was desirable to illustrate many more species if Indian ornithology was to be better served. Sálim Ali and S. Dillon Ripley attempted to do this in the 10 volumes of their *Handbook of the Birds of India and Pakistan* (1969-74) which contains colour illustrations of some 900 species of the 1200+ found in the Indian subcontinent, therefore still considerably short of the total. Moreover, the plates in the Handbook are spread over 10 separate volumes which makes the locating of individual species inconvenient and time-consuming since even in these plates the illustrations are not in systematic order, familywise. Under these conditions Bombay Natural History Society feels particularly gratified in its opportunity to publish, with the far-sighted sponsorship of U.S. Fish and Wildlife Service and Smithsonian Institution, Washington, this immaculately executed set of 106 plates (73 coloured, 33 black-and-white) by the well-known American bird painter John Henry Dick, depicting 1241 species of the Indian subcontinent. These include a few extralimital ones (EL) from just across our borders which could well be found within our territory at one time or another. Not only are the species systematically arranged in the plates, family by family, but in many cases both male and female as well as juvenile plumages and colour variants are also shown. The publication is particularly gratifying also because the sponsors intend it to be a non-profit distribution among low- and middle-income groups and among government and private scientific, cultural and educational institutions, and for the widest diffusion of knowledge concerning our rich ornithological heritage to enable well informed public participation in the conservation of wildlife and its habitats, and thus in the preservation of a healthy natural environment.

ACKNOWLEDGEMENTS

For various courtesies received by the artist, John Henry Dick, in the course of preparation of the plates, he wishes to thank the following institutions and individuals:

American Museum of Natural History, New York for making available their magnificent collection of Indian bird skins.

U.S. Fish and Wildlife Service (Department of Interior) for their financial assistance, their whole-hearted support and encouragement of this book from start to finish, and their belief that a complete illustrated volume on the birds of the Indian subcontinent would make a considerable contribution to scientific bird study and conservation as well as provide a handy reference for the general Indian public which is just awakening to the richness of its wildlife heritage.

Gerry (Gerard) Bertrand who was initially concerned with this project till his departure from FWS in 1980, and thereafter David Ferguson without whose active involvement in the catalytic role played by FWS throughout the protracted vicissitudes of the project it might never have materialized.

Ben King — who first conceived the idea of a complete illustrated guide to all the birds of greater

India. Thanks also for his original text and his careful selection of the AMNH skins; also for photographing rare specimens that could not leave the Museum.

Appreciation must also be accorded to the New York Zoological Society but for whose financial encouragement to Ben King in the initial stages this ambitious project might never have got off the ground.

On the Indian side, Bombay Natural History Society gratefully acknowledges the munificent gesture of the artist and the authors in waiving their claim to a royalty on this publication in favour of the Society. This they have done as a token of their admiration for the seminal work the Society has done in arousing public interest and awareness in Nature and Nature Conservation in India during the century of its existence. The useful suggestions and painstaking secretarial assistance rendered by Archna Mehrotra in the preparation of the captions and their coordination with the illustrations have helped greatly in expediting the work, and are thankfully acknowledged by the senior author.

EXPLANATIONS AND ABBREVIATIONS

The term SUBCONTINENT as used in this book includes the entire country south of the Himalayas — peninsular and continental India with its north-eastern extension into China and Burma. It also includes Pakistan, Nepal, Bhutan, Bangladesh and Sri Lanka. Overall, the area lies between latitudes 10° 30' and 40° N, and between longitudes 60° and 96°30'E. Some of the other terms and abbreviations employed in the captions also need to be spelled out. The excellent illustrations by John Henry Dick are sufficiently diagnostic by themselves not to need supplemental verbal descriptions. Omission of the latter, and the use of abbreviations, has enabled the adoption of a space-saving format — also more convenient for quick reference — of having the complete captions on the facing page of each plate. Of the numerals preceding the species-name in the caption, the first corresponds with the species number in *Handbook of the Birds of India and Pakistan* by Sálim Ali and S. Dillon Ripley (1969-74) for case of cross reference should more information on the species be desired. The second number (in brackets) identifies the illustration in the facing plate.

The following abbreviations have been used in the captions:

V = Vagrant (or occasional stray)
R = Resident. Also covers local migrants, i.e. birds that breed in one part of the Subcontinent in one season and move to other parts within the country in a different season, e.g. Pitta
M = Migrant (extralimital, chiefly long-distance)
RM= Resident with migratory populations (subspecies)
EL = Extralimital

ECLIPSE PLUMAGE. In certain families, e.g. Ducks (Anatidae), the males don a distinctive breeding plumage in summer and revert to a female-like (or eclipse) plumage in winter. The sexes are then difficult to distinguish. S (summer) and W (winter) against the illustration identify these seasonal plumages.

BIRDS OF PREY or RAPTORS often have very different and highly confusing plumages as adult, immature and juvenile. Moreover, in addition to the normal plumage they (and also some other birds) frequently have a dark phase (or morph) and a pale phase, with sometimes an intermediate one. All birds illustrated here are in adult plumage unless otherwise stated. The following abbreviations and symbols are used in the plates:

ad- = Subadult
imm = Immature
juv = Juvenile
S = Summer
W = Winter
br+ = Breeding
br- = Non-breeding
↓ = From above
↑ = From below
● = Dark phase
o = Pale phase
◉ = Intermediate phase

The word "Seasonally" before the numerals for altitude in the case of montane birds, signifies the limits within which the bird moves, the lower figure for winter, the higher for summer.

A comparison with some common and familiar bird as a standard to give a rough indication of size is helpful in field identification. Chiefly the following standards have been used in this book:

Sparrow	15 cm (6 in.)
Crow	42 cm (17 in.)
Quail	18-20 cm (7-8 in.)
Kite	60 cm (24 in.)
Bulbul	20 cm (8 in.)
Duck	60 cm (24 in.)
Myna	23 cm (9 in.)
Village hen	45-70 cm (18-28 in.)
Pigeon	32 cm (13 in.)
Vulture	90 cm (36 in.)
Partridge	32 cm (13 in.)

Plus (+) and minus (-) signs indicate whether the bird is bigger or smaller than the standard. Plus/minus (±) means that it is more or less the same size as the standard. The size in centimetres is also a rough approximation, and it must be treated as such.

HIMALAYAS (when used by itself) stands for the entire range west to east, including the 'command areas' along its base.

W HIMALAYAS extend from the westernmost limit of the range eastward to west-central Nepal and include the sections in Hazara, NWFP, Kashmir, Himachal Pradesh, Kumaon and Garhwal.

E HIMALAYAS extend from east-central Nepal eastward to NE Arunachal Pradesh and include the sections in Sikkim, Darjeeling district (W Bengal) and Bhutan.

NE HILL STATES include Meghalaya, Nagaland, Manipur, Mizoram, Tripura and, for our purpose, also the contiguous district of Chittagong in Bangladesh.

W GHATS COMPLEX stands for the entire range (Sahyadris), north to south, including the Nilgiri, Palni, Annamalai, and the associated hills of Karnataka, Tamil Nadu and Kerala.

E GHATS COMPLEX stands for the entire range north to south, including the Visakhapatnam ghats and the Nallamalai, Palkonda, Seshachalam, Shevaroy and associated hills of Orissa, Andhra Pradesh and Tamil Nadu.

CONTINENTAL INDIA. Upper India south of the Himalayas to northern Madhya Pradesh—chiefly the Indo-Gangetic Plain.

PENINSULAR INDIA. Roughly south of lat. 20° N (Tapti river).

BAY ISLANDS = The Andaman and Nicobar groups lying in the Bay of Bengal.

Terms used in the description of a bird's plumage and parts

Topography of a sparrow

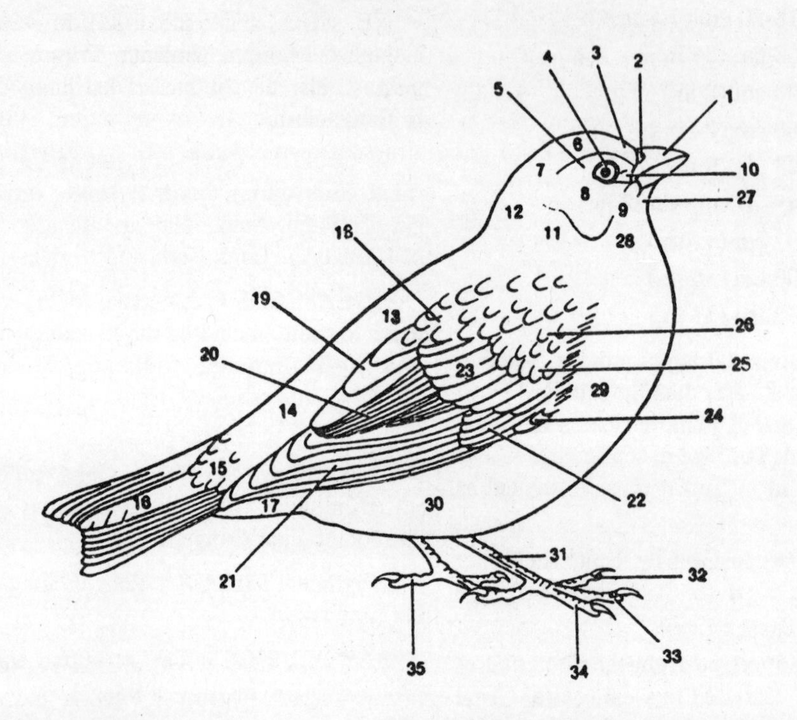

1	Culmen	18	Scapulars
2	Nostril	19	Tertials
3	Forehead	20	Secondaries } (remiges)
4	Iris	21	Primaries } (remiges)
5	Supercilium	22	Primary coverts
6	Crown	23	Greater coverts
7	Nape	24	Bastard wing (alula)
8	Ear-coverts	25	Median coverts
9	Malar region (malar stripe, moustache)	26	Lesser coverts
		27	Chin
10	Lores	28	Throat
11	Side of neck	29	Breast
12	Hindneck	30	Belly (abdomen)
13	Back	31	Tarsus
14	Rump	32	Inner toe
15	Upper tail-coverts	33	Middle toe
16	Tail (rectrices)	34	Outer toe
17	Under tail-coverts	35	Hind toe (hallux)

Systematic Index of Families, and Species in each

Family GAVIIDAE: Divers, Loons

Aquatic birds superficially resembling grebes but not very closely related to them. Toes fully webbed, not lobed or scalloped. Plumage dense, compact and harsh. Tarsi reticulate, laterally compressed. Legs short and set far back, almost at end. Wings short, narrow and tapering, set well back. Seen singly or in pairs. Frequent lakes, ponds and slow-flowing rivers in summer; essentially marine in winter. Skilled divers and swimmers. Take wing reluctantly and with some difficulty, but once airborne, their flight is powerful and swift. Sexes alike (winter); dimorphic (summer). Breeding: Extralimital.

		Plate *
1.+	BLACKTHROATED DIVER *Gavia arctica*	1(1)
2.	REDTHROATED DIVER *Gavia stellata*	1(2)

* The numeral in brackets following the plate number refers to the relevant species in the plate.
+ Number as in the Synopsis of the Birds of India and Pakistan by Dillon Ripley.

Family PODICIPITIDAE: Grebes

Aquatic birds with soft rudimentary tail, very small wings, and compressed sharply pointed bill. Legs placed far back, especially adapted for diving and swimming. Tarsi scutellated in front, laterally compressed. Front toes with broad lateral vane-like lobes. Hind toe small, raised, vertically lobed. Nails broad and flattened. Plumage dense and silky. Loath to fly, rise with effort, but once airborne can fly strongly, often long distances. Said to have a habit of eating own feathers and feeding them to the young, probably to aid digestion. Sexes seasonally dimorphic. Nest, a mass of water weeds with a central depression or floating mound of grass and rubbish, loosely anchored to reeds or the substrate. Incubation by both sexes. Downy young boldly striped blackish and white.

		Plate
3.	GREAT CRESTED GREBE *Podiceps cristatus*	1(3)
4.	BLACKNECKED GREBE *Podiceps nigricollis*	1(4)
4a.	REDNECKED GREBE *Podiceps griseigena*	1(5)
5.	LITTLE GREBE *Tachybaptus ruficollis*	1(6)

Family PROCELLARIIDAE: Petrels and Shearwaters

Sea birds of varying sizes and coloration, from myna to goose and of white, grey, brown, black plumage, or combinations of these. Bill short and stout to longish and slender, covered with horny plates, hooked at tip. Nostrils tubular. Wings narrow, long and pointed, with first primary longest and secondaries short. Tarsus short to medium, slender, laterally compressed, reticulated. Feet webbed, with strong hindclaw. Tail short, rounded. Sexes alike. Nest, a sand burrow excavated beneath scrub roots, near the shore. Incubation presumably by both sexes.

		Plate
6.	CAPE PETREL *Daption capensis*	2(12)
8.	PINKFOOTED SHEARWATER *Procellaria carneipes*	2(8)
9.	WEDGETAILED SHEARWATER *Procellaria pacifica*	2(7)
11.	AUDUBON'S SHEARWATER *Procellaria lherminieri*	2(13)
13a.	JOUANIN'S GADFLY PETREL *Bulweria fallax*	2(6)
13b.	BULWER'S GADFLY PETREL *Bulweria bulwerii*	2(5)
EL.	STREAKED SHEARWATER *Calonestris leucomelas*	2(10)
EL.	SHORT-TAILED SHEARWATER *Procellaria tenuirostris*	2(9)

Family HYDROBATIDAE: Storm Petrels

The smallest sea birds, from about sparrow to myna in size, of blackish or greyish plumage, mostly with a white rump. Wings long; tail medium to long; neck short. Bill slender, of medium length, grooved, hooked at tip. Nostrils tubular with a single orifice. Legs slender, medium to long; feet webbed, mostly black. Webs black or particoloured. Feed by 'walking' or 'hopping' on the water with wings fluttering and held slightly above line of back, long legs dangling, feet paddling, head bent low and bill touching the surface. Sexes alike. Nest in crevices in cliffs.

		Plate
14.	WILDON'S STORM PETREL *Oceanites oceanicus*	2(1)
15.	DUSKYVENTED STORK PETREL *Fregetta tropica*	2(4)
16.	LEACHE'S OR FORKTAILED STORM PETREL *Oceanodroma leucorhoa*	2(2)
EL.	WHITEFACED STORM PETREL *Pelagodroma marine*	2(11)
EL.	MATSUDAIRA'S STORM PETREL *Oceanodroma matsudàirae*	2(3)

Family PHAETHONTIDAE: Tropic-birds

Tropical sea birds superficially resembling terns but morphologically closer to cormorants and frigate birds. Plumage chiefly white and black. Head large; neck short; bill yellow or orange-red, longish, stout, decurved and pointed. Wings long and pointed. Tail wedge-shaped with the middle pair of feathers in adults narrow, ribbon-like and much elongated. Legs short; feet webbed. Sexes alike. Nest, under shelter of a ledge of rock or in a crevice. A single egg is laid, curiously like that of raptors. Young hatch covered with down.

		Plate
17.	REDBILLED (OR SHORT-TAILED) TROPIC-BIRD *Phaethon aethereus*	5(2)
18.	REDTAILED TROPIC-BIRD *Phaethon rubricauda*	5(1)
19.	LONGTAILED TROPIC-BIRD *Phaethon lepturus*	5(4)

Family PELECANIDAE: Pelicans

Large, gregarious, squat and clumsy fish-eating birds with short sturdy legs and large webbed feet. Tarsus compressed and reticulate in front. Wings large and broad; tail short, square and soft. Characteristic of the Family is the long heavy bill with the upper mandible flattened and hook-tipped and the lower consisting of two narrow flexible arches underhung throughout its length by a capacious gular pouch of naked skin which functions as a drag net for scooping up fish while swimming. Flight is strong and over long distances, though the initial take-off requires some effort. They fly with the neck bent back in a flat S, head drawn in between the shoulders. Pelicans fly either in the characteristic V-shaped echelons of geese, or in long straggly ribbons with a wide front. Often seen soaring on thermals. Sexes alike. Nesting in colonies. Nest of sticks in trees, or on reeds and grass or mud on the ground. Eggs 2-4. Chicks naked when newly hatched.

		Plate
20.	ROSY or WHITE PELICAN *Pelecanus onocrotalus*	3(2)
21.	SPOTTEDBILLED PELICAN *Pelecanus philippensis*	3(1)
22.	DALMATIAN PELICAN *Pelecanus philippensis crispus*	3(3)

Family SULIDAE: Boobies

Sea birds, typically white with variable amount of black in wings. Body stout, neck of moderate length, wings long and pointed, tail rather long and wedge-shaped. Bill stout, conical, pointed, slightly downcurved at tip, but not hooked. Bill, and bare facial and gular skin, brightly coloured. Legs short and stout, feet large and fully webbed. They have a characteristic habit of plunging headlong into the water from a height of 8 to 17 metres in the air for fish, and are capable of very deep dives in underwater pursuit of fish. Breed on oceanic islands, on the ground or in stick nest in trees.

		Plate
23.	MASKED BOOBY *Sula dactylatra*	5(3)
24.	REDFOOTED BOOBY *Sula sula*	5(5)
25.	BROWN BOOBY *Sula leucogaster*	5(6)

FAMILY PHALACROCORACIDAE: Cormorants and Darter

Gregarious, fish-eating, colonial-nesting water birds. All Indian species characterized by black plumage. Bill laterally compressed, rather long, slender and pointed; hooked at tip in cormorants, stiletto-like in darters. Neck and body long, spindle-shaped; wings of moderate length. Tail long and stiff. Legs short; feet large and webbed; claws much curved. Plumage not very dense or resistant to water; becomes soaked by prolonged immersion and requires constant drying out. The birds rise off water with some difficulty, but the flight is powerful and sustained, with the neck stretched out in front. Cormorants fly in goose-like V-shaped echelons or wavy diagonal lines. Sexes alike. Nests, skimpy to fairly substantial stick platforms lined with water weeds in trees, rarely on rocks. Eggs 2-6. Incubation by both sexes. Young nidicolous, naked at hatching, down-covered later, and extraordinarily ugly.

		Plate
26.	CORMORANT *Phalacrocorax carbo*	4(1)
27.	INDIAN SHAG *Phalacrocorax fuscicollis*	4(2)
28.	LITTLE CORMORANT *Phalacrocorax niger*	4(3)
28a.	PYGMY CORMORANT *Phalacrocorax pygmaeus*	4(4)
29.	DARTER *Anhinga rufa*	4(5)

Family FREGATIDAE: Frigate Birds or Man-of-war Hawks

Large, gregarious, dark coloured or black-and-white oceanic birds with long, pointed, streamlined wings, deeply forked tails, and flight resembling that of raptors. Bill long and strongly hooked, rounded in cross-section, the culmen convex. Throat bare, bright coloured gular pouch occasionally inflated to ridiculous proportions by courting male, even in flight. Tarsus very short, stout, feathered. Feet small and webbed at the base; claws long, strong, much curved; middle claw pectinate. Usually feed by pirating food obtained by other birds. Frigate birds are magnificent fliers and capable of soaring and sailing for hours, spending almost all day on the wing. Nest of sticks built on trees and bushes and on rocks. Eggs 1 or 2. Incubation by both sexes. Chick blind and naked at hatching; covered with white down later.

		Plate
31.	LESSER FRIGATE BIRD *Fregata minor*	3(5)
32.	LEAST FRIGATE BIRD *Fregata ariel*	3(4)

Family ARDEIDAE: Herons, Egrets, Bitterns

Long-legged, lanky wading birds, with long slender flexible necks which are retracted into a flat S during flight. Bill long, straight, sharp-pointed and dagger-like. Tarsi very long; toes long and slender, the middle and outer toes united by a small web at their base; claw of middle toe pectinate. Most species have curious powder-down patches on each side of rump and breast providing a sort of dry shampoo for degreasing soiled feathers. Plumage soft and loose-textured, usually white, grey, purple or brown. In many species filamentous ornamental plumes acquired during the breeding season, for the trade in which the birds were greatly persecuted and in some places almost exterminated. Sexes alike or nearly so. Breeding, colonial, usually in mixed heronries. Nest, shallow stick platform on trees. Eggs 3-6. Young nidicolous.

		Plate
33.	GREAT WHITEBELLIED HERON *Ardea insignis*	6(3)
34.	GIANT HERON *Ardea goliath*	6(2)
36.	GREY HERON *Ardea cinerea*	6(4)
37.	PURPLE HERON *Ardea purpurea*	6(5)
38.	LITTLE GREEN HERON *Ardeola striatus*	6(1)
42.	POND HERON OR PADDYBIRD *Ardeola grayii*	4(9), 6(8)
43.	CHINESE POND HERON *Ardeola bacchus*	6(7)
44.	CATTLE EGRET *Bubulcus ibis*	7(5)
46.	LARGE EGRET *Ardea alba*	7(1)
47.	SMALLER EGRET *Egretta intermedia*	7(2)
49.	LITTLE EGRET *Egretta garzetta*	4(7), 7(3)
50.	INDIAN REEF HERON *Egretta gularis*	7(4)
52.	NIGHT HERON *Nycticorax nycticorax*	4(10), 6(14)
53.	MALAY or TIGER BITTERN *Grasachius melanolophus*	4(8), 6(15)
55.	LITTLE BITTERN *Ixobrychus minutus*	6(11)
56.	CHESTNUT BITTERN *Ixobrychus cinnamomeus*	6(9)
57.	YELLOW BITTERN *Ixobrychus sinensis*	6(10)

		Plate
58.	BLACK BITTERN *Ixobrychus flavicollis*	6(12)
59.	BITTERN *Botaurus stellaris*	4(6), 6(13)
EL.	SQUACCO HERON *Ardeola ralloides*	6(6)

Family CICONIIDAE: Storks

Large, long-legged, diurnal birds chiefly terrestrial and marsh-haunting. Colour pattern mainly white and black with a metallic sheen. Bill long, massive, pointed straight or nearly so, and ungrooved. Wings long and broad. Tail short; under tail-coverts lax and greatly developed in some species. Legs very long, tibiae partly naked; toes of moderate length, webbed at base, all four at same level (*contra* cranes which have the hind toe raised above the ground); claws blunt. Lacking voice muscles therefore vocally silent. Most species produce low grunting and hissing noises, and a loud castanet-like clattering or snapping of the mandibles. Storks are strong fliers, flying with neck and legs fully outstretched. Nest, large stick platforms in trees, or on cliffs and buildings. Eggs 3-6. Incubation and nest-feeding by both sexes. Chicks nidicolous; naked at first, downy later. Food regurgitated by parent into nest, whence guzzled by nestlings.

	Plate
60. PAINTED STORK *Mycteria leucocephala*	8(2)
61. OPENBILL STORK *Anastomus oscitans*	8(5)
62. WHITENECKED STORK *Ciconia episcopus*	8(6)
63. WHITE STOCK *Ciconia ciconia*	8(3)
65. BLACK STORK *Ciconia nigra*	8(4)
66. BLACKNECKED STORK *Ephippiorhynchus asiaticus*	8(1)
67. ADJUTANT *Leptoptilos dubius*	8(10)
68. LESSER ADJUTANT *Leptoptilos javanicus*	8(12)

Family THRESKIORNITHIDAE: Ibises, Spoonbill

Comparatively long and bare-legged gregarious waterside or marsh birds related to and resembling storks, herons, and egrets, with whom they normally associate. Plumage chiefly white or black, or chestnut with metallic gloss. Bill long, slender, grooved and decurved (ibises), or extremely flattened and spatulate at tip (spoonbill). Face and throat, or whole head and neck bare in some species. Neck slender, longish, outstretched in flight. Wings long; tail short. Legs and toes moderately long; tibiae partly bare; toes webbed at base. Sexes nearly alike. Fly in V-formation or in wavy diagonal ribbons. Perch and roost on trees. Nest, a platform of sticks on trees. Eggs 2-4, rarely 5. Incubation by both sexes.

		Plate
69.	WHITE IBIS *Threskiornis aethiopica*	8(11)
70.	BLACK IBIS *Pseudibis papillosa*	8(7)
71.	GLOSSY IBIS *Plegadis falcinellus*	8(9)
72.	SPOONBILL *Platalea leucorodia*	8(8)

Family PHOENICOPTERIDAE: Flamingos

Large, excessively long-legged marsh birds with very long slender necks and peculiar thick lamellate bills sharply downcurved or 'broken' in the middle. Plumage largely pinkish white and crimson, with black remiges or wing-quills. Tibiae bare; toes short and webbed. Highly gregarious, often in vast congregations numbering over many hundred thousand. Feed in shallow, usually brackish water with head immersed. Sometimes swimming and 'up-ending' in deeper water. The partly open bill is inverted so that the upper mandible forms a scoop with the culmen skimming or scraping the bottom ooze. The fleshy tongue works back and forth like a piston sucking in the water and mud, from which minute organisms are strained out by the lamellae along the edges of the bill. Flight, with fairly rapid wing-strokes, neck fully extended in front and legs trailing well behind, in diagonal wavy ribbons, single file, or in V-formation. Breeding in colonies, with hundreds of nests close to one another in a compact, expansive 'city' covering several acres. Nest, a truncated conical mound with shallow pan-like depression at top, from a few centimetres to half a metre high, of sun-baked mud scraped from the vicinity when in semi-liquid condition, and daubed on; sometimes merely a raised bed of mud pellets. Eggs normally 1, occasionally 2. Incubation by both sexes. Newly hatched, and small chicks, fed by parent on drops of a clear liquid of unknown origin and composition from its bill tip.

		Plate
73.	FLAMINGO *Phoenicopterus roseus*	3(7)
74.	LESSER FLAMINGO *Phoeniconaias minor*	3(6)

Family ANATIDAE: Ducks, Geese, Swans

Large conspicuous wetland birds, better known generally than perhaps any other group, and hunted universally. Considerable diversity in size and coloration — from pigeon to vulture, and from wholly white to combinations of grey, brown, black and green with metallic reflections. Bill typically broad, flat, rounded at tip, and with a comb-like fringe or lamellae for straining out food particles from water, in which they chiefly feed. Wings mostly rather narrow and pointed, adapted for swift and long-ranging flight. Tail short. Legs short; feet webbed. Most species are migratory. Nest on the ground, or in holes or hollows in tree trunks, etc. Eggs 6-16. Young nidifugous, down covered.

		Plate
75.	SIBERIAN REDBREASTED GOOSE *Branta ruficollis*	12(7)
76.	BEAN GOOSE *Anser fabalis*	12(9)
79.	WHITEFRONTED GOOSE *Anser albifrons*	12(10)
80.	LESSER WHITEFRONTED GOOSE *Anser erythropus*	12(8)
81.	GREYLAG GOOSE *Anser anser*	12(11)
82.	BARHEADED GOOSE *Anser indicus*	12(12)
84.	BEWICK'S SWAN *Cygnus cygnus bewickii*	3(10)
86.	WHOOPER SWAN *Cygnus cygnus*	3(8)
87.	MUTE SWAN *Cygnus olor*	3(9)
88.	LESSER WHISTLING TEAL or TREE DUCK *Dendrocygna javanica*	11(4), 13(4)
89.	LARGE WHISTLING TEAL *Dendrocygna bicolor*	11(3) 13(3)
90.	RUDDY SHELDUCK *Tadorna ferruginea*	11(1), 13(1)
91.	COMMON SHELDUCK *Tadorna tadorna*	11(2), 13(2)
92.	MARBLED TEAL *Marmaronetta angustirostris*	9(8), 10(8)
93.	PINTAII *Anas acuta*	9(2), 10(2)
94.	COMMON TEAL *Anas crecca*	9(9), 10(9)
95.	BAIKAL TEAL *Anas formosa*	9(7), 10(7)

		Plate
96.	GREY TEAL *Anas gibberifrons*	9(5), 10(5)
97.	SPOTBILL DUCK *Anas poecilorhyncha*	9(3), 10(3)
100.	MALLARD *Anas platyrhynchos*	9(1), 10(1)
101.	GADWALL *Anas strepera*	9(12),10(12)
102.	FALCATED TEAL *Anas falcata*	9(11), 10(11)
103.	WIGEON *Anas penelope*	9(4), 10(4)
104.	GARGANEY *Anas querquedula*	9(10), 10(10)
105.	SHOVELLER *Anas clypeata*	9(6), 10(6)
106.	PINKHEADED DUCK *Rhodonessa caryophyllacea*	11(13)
107.	REDCRESTED POCHARD *Netta rufina*	11(5), 13(5)
108.	COMMON POCHARD *Aythya ferina*	11(6), 13(6)
109.	WHITE-EYED POCHARD OR FERRUGINOUS DUCK *Aythya nyroca*	11(10), 13(8)
110.	BAER'S POCHARD *Aythya baeri*	11(9)
111.	TUFTED DUCK *Aythya fuligula*	11(7), 13(7)
112.	SCAUP DUCK *Aythya marila*	11(8), 13(9)
114.	COTTON TEAL OR QUACKY-DUCK *Nettapus coromandelianus*	11(12), 13(12)
115.	COMB DUCK *Sarkidiornis melanotos*	12(5)
116.	WHITEWINGED WOOD DUCK *Cairina scutulata*	12(6)
117.	LONGTAIL OR OLD SQUAW DUCK *Clangula hyemalis*	12(3)
118.	GOLDENEYE DUCK *Bucephala clangula.*	12(2)
119.	SMEW *Mergus albellus*	11(11), 13(10)
120. 121.	GOOSANDER, COMMON MERGANSER *Mergus merganser*	12(4), 13(11)
123.	WHITEHEADED STIFFTAILED DUCK *Oxyura leucocephala*	12(1), 13(13)

Family ACCIPITRIDAE: Hawks, Vultures, etc.

The Raptors or diurnal Birds of Prey are a major component of the order Falconiformes. Bill short with upper mandible longer than lower, curved and strongly hooked at the tip; basal portion covered with a cere which is usually bright coloured. Wings rounded. Feet strong; tarsi normally partly or fully feathered; hallux always present; claws hooked and powerful. Many species have confusingly different adult and juvenile plumages. Sexes nearly alike; female usually larger. Feed on the flesh of animals, self-killed or carrion. Breed in trees, or on crags. Nest made of sticks and often lined with leaves, grass, etc. Eggs 1-5. The rate of reproduction, especially in the larger species, is slow.

		Plate
124.	BLACKWINGED or BLACKSHOULDERED KITE *Elanus caeruleus*	22(1),23(3), 28(2)
125.	BLYTH'S BAZA *Aviceda jerdoni*	22(3), 21(8), 23(2)
127.	BLACKCRESTED BAZA *Aviceda leuphotes*	22(2), 23(1), 28(1)
130.	HONEY BUZZARD *Pernis ptilorhyncus*	24(1), 25(1)
131.	RED KITE *Milvus milvus*	14(10), 15(1), 17(3)
132.	BLACK KITE *Milvus migrans migrans*	14(8), 15(4)
133.	PARIAH KITE *Milvus migrans govinda*	14(7), 15(3), 17(1)
134.	BLACKEARED or LARGE INDIAN KITE *Milvus migrans lineatus*	14(9), 15(2)

		Plate
135.	BRAHMINY KITE *Haliastur indus*	14(11), 17(2), 25(2)
136.	GOSHAWK *Accipiter gentilis*	22(10), 23(11)
139.	SHIKRA *Accipter badius*	22(4), 23(5)
141.	CAR NICOBAR SHIKRA *Accipiter badius butleri*	22(7), 23(8)
143.	HORSFIELD'S GOSHAWK *Accipiter soloensis*	22(6), 23(6)
144.	CRESTED GOSHAWK *Accipiter trivirgatus*	22(5), 23(4)
147.	SPARROW-HAWK *Accipiter nisus nisosimilis*	22(11) 23(10)
148.	SPARROW-HAWK *Accipiter nisus melaschistos*	22(12), 23(10)
151.	BESRA SPARROW-HAWK *Accipiter virgatus besra*	22(8), 23(9)
152.	EASTERN BESRA SPARROW-HAWK *Accipiter virgatus gularis*	22(9), 23(7)
153.	LONGLEGGED BUZZARD *Buteo rufinus*	24(6), 25(8)
154.	UPLAND BUZZARD *Buteo hemilasius*	24(5), 25(9)
155.	DESERT BUZZARD *Buteo buteo vulpinus*	24(3), 25(4)
156.	JAPANESE BUZZARD *Buteo buteo japonicus*	24(2), 25(5)
157.	WHITE-EYED BUZZARD-EAGLE *Butastur teesa*	25(3), 28(3), 29(3)
158.	HODGSON'S HAWK-EAGLE *Spizaetus nipalensis*	20(4), 21(10)
160.	CHANGEABLE HAWK-EAGLE *Spizaetus cirrhatus limnaeetus*	20(5), 21(9)
161.	CRESTED HAWK-EAGLE *Spizaetus cirrhatus cirrhatus*	20(6), 21(7), 28(12)
163.	BONELLI'S EAGLE *Hieraaetus fasciatus*	20(1), 21(2), 28(11)
164.	BOOTED HAWK-EAGLE *Hieraaetus pennatus*	20(2), 17(5), 25(11)
165.	RUFOUSBELLIED HAWK-EAGLE *Hieraaetus kienerii*	20(3), 25(10)
166.	GOLDEN EAGLE *Aquila chrysaetos*	26(6), 27(6), 28(9)
167.	IMPERIAL EAGLE *Aquila heliaca*	26(7), 27(7), 28(10)
168.	TAWNY EAGLE *Aquila rapax vindhiana*	26(4) 27(5), 28(8)
169.	EASTERN STEPPE EAGLE *Aquila rapax nipalensis*	26(5), 27(4), 28(6)
170.	GREATER SPOTTED EAGLE *Aquila clanga*	26(3), 27(3), 28(7)
171.	LESSER SPOTTED EAGLE *Aquila pomarina*	26(2), 27(2)
172.	BLACK EAGLE *Ictinaetus malayensis*	26(1), 27(1)
172a.	WHITETAILED EAGLE *Haliaeetus albicilla*	14(4), 15(8)
173.	WHITEBELLIED SEA EAGLE *Haliaeetus leucogaster*	14(1), 15(6)
174.	PALLAS'S FISHING EAGLE *Haliaeetus leucoryphus*	14(5), 15(9)
175.	GREYHEADED FISHING EAGLE *Ichthyophaga ichthyaetus*	14(2), 15(7)
177.	HIMALAYAN GREYHEADED FISHING EAGLE *Ichthyophaga nana*	14(3), 15(5)
178.	BLACK OR KING VULTURE *Sarcogyps calvus*	16(7), 17(8), 18(4)
179.	CINEREOUS VULTURE *Aegypius monachus*	16(6), 17(7), 18(3)
180.	GRIFFON VULTURE *Gyps fulvus*	16(5), 17(10), 18(8)
181.	HIMALAYAN GRIFFON *Gyps himalayensis*	16(9), 17(9), 18(6)
182.	INDIAN LONGBILLED VULTURE *Gyps indicus*	16(4), 17(12), 18(7)
185.	INDIAN WHITEBACKED VULTURE *Gyps bengalensis*	16(3), 17(11), 18(5)
186.	EGYPTIAN OR SCAVENGER VULTURE *Neophron percnopterus*	16(1), 17(4), 18(1)
188.	BEARDED VULTURE or LAMMERGEIER *Gypaetus barbatus*	16(2), 17(6), 18(2)
189.	HEN-HARRIER *Circus cyaneus*	19(1)
190.	PALE HARRIER *Cirus macrourus*	14(13), 19(2)
191.	MONTAGU'S HARRIER *Circus pygargus*	19(3)
192.	PIED HARRIER *Circus melanoleucos*	14(12), 19(4)
193.	MARSH HARRIER *Circus aeruginosus*	14(14), 19(5)

		Plate
194.	EASTERN MARSH HARRIER (or STRIPED HARRIER) *Circus aeruginosus spilonotus*	19(6)
195.	SHORT-TOED EAGLE *Circaetus gallicus*	20(7), 21(3)
196.	CRESTED SERPENT EAGLE *Spilornis cheela*	20(8), 21(4) 28(4)
200.	ANDAMAN PALE SERPENT EAGLE *Spilornis cheela davisoni*	20(9)
201.	NICOBAR CRESTED SERPENT EAGLE *Spilornis cheela minimus*	20(10)
202.	GREAT NICOBAR SERPENT EAGLE *Spilornis klossi*	20(11), 21(5)
202a.	ANDAMAN DARK SERPENT EAGLE *Spilornis elgini*	20(12), 21(6)
203.	OSPREY *Pandion haliaetus*	14(6), 21(1)
EL.	ROUGHLEGGED BUZZARD *Buteo lagopus*	24(4), 25(7), 28(5)

Family FALCONIDAE: Falcons

Diurnal birds of prey, with grey or brown and white or buff plumage; streaked or barred, especially below, showing strong contrast of black and white. Bill short, strongly hooked and toothed. Wings long and pointed. Legs strong; toes long; claws hooked and powerful. Extremely accomplished and swift fliers, generally killing their flying prey on the wing. Food chiefly birds, large insects, rodents and other small ground animals. Some species extensively used in the sport of Falconry. There is considerable change in the plumage from immature to adult which makes identification in the field and from descriptions difficult and unsatisfactory. Sexes alike, but female larger. Normally old stick nests of other birds in trees, or in cliffs are appropriated. Eggs, 2-5, usually 3 or 4.

		Plate
204.	REDBREASTED FALCONET *Microhierax caerulescens*	29(2), 30(3)
205.	WHITELEGGED FALCONET *Microhierax melanoleucos*	29(1), 30(1)
206.	SAKER or CHERRUG FALCON *Falco biarmicus cherrug*	29(16), 30(11)
207.	SHANGHAR FALCON *Falco biarmicus milvipes*	29(17)
208.	LAGGAR FALCON *Falco biarmicus jugger*	29(14), 30(14)
209.	PEREGRINE FALCON *Falco peregrinus japonensis*	29(12), 30(13)
211.	SHAHEEN FALCON*Falcoperegrinusperegrinator*	29(13), 30(12)
212.	HOBBY *Falco subbuteo*	29(7), 30(10)
215.	ORIENTAL HOBBY *Falco severus*	29(10), 30(9)
216.	SOOTY FALCON *Falco concolor*	29(11), 30(8)
217.	MERLIN *Falco columbarius*	29(9), 30(6)
219.	REDHEADED MERLIN *Falco chicquera*	29(8), 30(7)
220.	REDLEGGED FALCON *Falco vespertinus*	29(5), 30(5)
221.	LESSER KESTREL *Falco naumanni*	29(4), 30(2)
222.	KESTREL *Falco tinnunculus*	29(6), 30(4)
EL.	BARBARY FALCON *Falco pelegrinoides*	29(15)

Family MEGAPODIIDAE: Megapodes

Birds closely related to the pheasants, junglefowl, etc. Bill moderate. Legs and feet very large and powerful; tarsus in front broadly scutellated; claws long and straight; hindclaw longest. Wings and tail short and rounded. Sexes alike. The Family is remarkable for the reptile-like nesting habits, laying their eggs in holes in the ground or in scraped-together mounds of sand and humus, leaving them to hatch by the heat of the sun and/or decomposition of the vegetable matter. Young born with full juvenile plumage and able to fly and fend for themselves. Chiefly confined to the Australasian Region; represented within our limits by a single species.

	Plate
225. MEGAPODE *Megapodius freycinet*	32(18)

Family PHASIANIDAE: Pheasants, Partridges, Quails, etc.

Comprises the "Game Birds" which form a valuable food resource for man. Terrestrial, but may roost on trees. Bill thick and short with the upper mandible overhanging the lower. Legs stout and unfeathered, usually armed with one or more pointed spurs in male; hallux always present; claws short, blunt and very strong for scratching the ground for food. Wings short and rounded. Flight may be swift and strong, but not for long distances. Normally feed on grain, seeds and tender shoots; also fruits, insects, etc. The majority lay their eggs (4-8, sometimes more) on the ground in open scrapes with no or scanty lining. Young nidifugous and downy.

		Plate
227.	SNOW PARTRIDGE *Lerwa lerwa*	32(14)
228.	SEESEE PARTRIDGE *Ammoperdix griseogularis*	32(13)
229.	TIBETAN SNOWCOCK *Tetraogallus tibetanus*	34(8)
232.	HIMALAYAN SNOWCOCK *Tetraogallus himalayensis*	34(9)
233.	PHEASANT-GROUSE *Tetraophasis szechenyii*	34(3)
236.	CHUKAR PARTRIDGE *Alectoris chukar*	32(16)
238.	BLACK PARTRIDGE *Francolinus francolinus*	31(1)
241.	PAINTED PARTRIDGE *Francolinus pictus*	31(4)
243.	CHINESE FRANCOLIN *Francolinus pintadeanus*	31(3)
246.	GREY PARTRIDGE *Francolinus pondicerianus*	31(2)
247.	SWAMP PARTRIDGE or KYAH *Francolinus gularis*	31(5)
249.	TIBETAN PARTRIDGE *Perdix hodgsoniae*	32(17)
250.	GREY QUAIL *Coturnix coturnix*	32(6)
252.	BLACKBREASTED or RAIN QUAIL *Coturnix coromandelica*	32(8)
253.	BLUEBREASTED QUAIL *Coturnix chinensis*	32(4)
255.	JUNGLE BUSH QUAIL *Perdicula asiatica*	32(11)
261.	ROCK BUSH QUAIL *Perdicula argoondah*	32(10)
263.	PAINTED BUSH QUAIL *Perdicula erythrorhyncha*	32(7)
265.	MANIPUR BUSH QUAIL *Perdicula manipurensis*	32(9)
267.	COMMON HILL PARTRIDGE *Arborophila torqueola*	31(6)
270.	RUFOUSTHROATED HILL PARTRIDGE *Arborophila rufogularis*	31(9)
271.	Ssp. *intermedia* of 270	31(10)
272.	WHITECHEEKED HILL PARTRIDGE *Arborophila atrogularis*	31(7)
273.	REDBREASTED HILL PARTRIDGE *Arborophila mandellii*	31(8)
274.	BAMBOO PARTRIDGE *Bambusicola fytchii*	32(15)
275.	RED SPURFOWL *Galloperdix spadicea*	31(11)
278.	PAINTED SPURFOWL *Galloperdix lunulata*	31(13)

		Plate
279.	CEYLON SPURFOWL *Galloperdix bicalcarata*	31(12)
280.	MOUNTAIN QUAIL *Ophrysia superciliosa*	32(12)
282.	BLOOD PHEASANT *Ithaginis cruentus*	35(5)
285.	WESTERN TRAGOPAN *Tragopan melanocephalus*	34(1)
286.	CRIMSON or SATYR TRAGOPAN *Tragopan satyra*	34(2)
288.	BLYTH'S or GREYBELLIED TRAGOPAN *Tragopan blythii*	34(5)
289.	TEMMINCK'S TRAGOPAN *Tragopan temminckii*	34(4)
290.	IMPEYAN or MONAL PHEASANT *Lophophorus impejanus*	34(6)
291.	SCLATER'S MONAL *Lophophorus sclateri*	34(7)
292.	EARED PHEASANT *Crossoptilon crossoptilon*	34(10)
294.	KALEEJ PHEASANT *Lophura leucomelana*	35(4), 105(2)
295.	Ssp. *melanota* of 294	105(3)
296.	Ssp. *lathami* of 294	105(4)
297.	Ssp. *williamsi* of 294	105(1)
298.	Ssp. *moffitti* of 294	105(5)
299.	RED JUNGLEFOWL *Gallus gallus*	35(6)
301.	GREY or SONNERAT'S JUNGLEFOWL *Gallus sonnerattii*	35(9)
302.	CEYLON JUNGLEFOWL *Gallus lafayetii*	35(8)
304.	KOKLAS PHEASANT *Pucrasia macrolopha*	35(1)
307.	CHIR PHEASANT *Catreus wallichii*	35(2)
308.	MRS HUME'S BARREDBACK PHEASANT *Syrmaticus humiae*	35(7)
310.	PEACOCK-PHEASANT *Polyplectron bicalcaratum*	35(10)
311.	COMMON PEAFOWL *Pavo cristatus*	33(1)
312.	BURMESE PEAFOWL *Pavo muticus*	33(2)
EL.	COMMON PHEASANT *Phasianus colchicus*	35(3)

Family TURNICIDAE: Bustard-Quails

Small terrestrial birds superficially very like the true quails. But the hallux (hind toe) is absent making the feet three-toed as in bustards. Poor fliers. Females larger than males, more showily coloured, polyandrous, and with a more active role in courtship. Nest, a scrape or depression in the ground sometimes lined with grass. Eggs, usually 4. Incubation, and rearing of the chicks done entirely by the male. Young down-covered and precocial.

		Plate
313.	LITTLE BUSTARD-QUAIL *Turnix sylvatica*	32(1)
314.	YELLOWLEGGED BUTTON QUAIL *Turnix tanki*	32(5)
318.	COMMON BUSTARD-QUAIL *Turnix suscitator*	32(2)
319.	Ssp. *leggei* of 318.	32(3)

Family GRUIDAE: Cranes

Large, longlegged, terrestrial birds superficially resembling storks. Bill pointed and comparatively short. Tibiae partly bare; toes short, strong, unwebbed; hind toe raised above level of the other three (*contra* Storks). Wings large and broad with the inner secondaries elongated, curled and drooping (used in display) under which the short tail is hidden. Flight powerful, in V-formation or in line. Sexes nearly alike.

Cranes pair for life; they have exceptionally powerful, resonant, bugle-like calls, and a spectacular ceremonial 'dance' which consists of curtseying, prancing, capering and wildly leaping at, around, and away from each other, accompanied by noisy duetting. Food, vegetable matter, grain, insects, small animals, etc. Nest, a mass of vegetation on the ground or in shallow water. Eggs, usually 2.

		Plate
320.	COMMON CRANE *Grus grus*	7(10)
321.	BLACKNECKED CRANE *Grus nigricollis*	7(9)
322.	HOODED CRANE *Grus monacha*	7(11)
323.	SARUS CRANE *Grus antigone*	7(7)
325.	SIBERIAN CRANE *Grus leucogeranus*	7(6)
326.	DEMOISELLE CRANE *Anthropoides virgo*	7(8)

Family RALLIDAE: Rails, Coots

Small to medium-sized marsh or water birds, with some terrestrial forms. Plumage black or in soft shades of grey, brown, blue or green. Body laterally compressed. Bill strong; stout and short or curved and long. Legs and toes long; tibiae partly bare. Wings short and rounded. Generally poor fliers, with some island forms nearly flightless; however, some species are known to make long migratory flights. Sexes nearly alike. Feed on all types of animal and vegetable matter. Nest, on the ground, in bushes, or floating on water. Eggs, 2-16. Incubation by both sexes.

		Plate
327.	WATER RAIL *Rallus aquaticus*	36(1)
329.	BLUEBREASTED BANDED RAIL *Rallus striatus*	36(2)
331.	REDLEGGED BANDED CRAKE *Rallina fasciata*	36(3)
332.	BANDED CRAKE *Rallina eurizonoides*	36(5)
333.	ANDAMAN BANDED CRAKE *Rallina canningi*	36(12)
334.	CORN CRAKE *Crex crex*	36(7)
335.	LITTLE CRAKE *Porzana parva*	36(4)
337.	BAILLON'S CRAKE *Porzana pusilla*	36(6)
338.	SPOTTED CRAKE *Porzana porzana*	36(8)
339.	RUDDY CRAKE *Porzana fusca*	36(9)
341.	ELWES'S CRAKE *Amaurornis bicolor*	36(11)
342.	BROWN CRAKE *Amaurornis akool*	36(10)
343.	WHITEBREASTED WATERHEN *Amaurornis phoenicurus*	36(17)
346.	WATER COCK *Gallicrex cinerea*	36(13)
347.	MOORHEN *Gallinula chloropus*	36(14)
349.	PURPLE MOORHEN *Porphyrio porphyrio*	36(19)
350.	COOT *Fulica atra*	36(15)

Family HELIORNITHIDAE: Finfoots

Shy and secretive coot-like birds inhabiting dense swampy forest. Legs short and very strong; toes fringed with a lobed web. Wings rounded; tail broad and stiff. Flight strong, but low and not prolonged. Sexes nearly alike. Nest, a pad of twigs on trees over water. Eggs, 2-6. Incubation by both sexes. A single Indomalayan species.

		Plate
351.	MASKED FINFOOT *Heliopais personata*	1(7)

Family OTIDIDAE: Bustards

Large terrestrial birds inhabiting open grassy plains. Plumage buff, grey, white and black, generally cryptically patterned. Neck long. Bill short, stout and flattened. Legs strong and long; tibiae bare; foot cursorial, with three short and broad anterior toes. Wings broad. Flight strong, but not very frequent, the birds preferring to run or crouch when faced with danger. Sexes dimorphic. Food, both vegetable and animal matter. Nest, a scrape on bare ground, sometimes lined with grass, etc., or under a bush. Eggs 1-5. Incubation by female. Young, nidicolous and downy.

		Plate
352.	GREAT BUSTARD *Otis tarda*	37(2)
353.	LITTLE BUSTARD *Otis tetrax*	37(1)
354.	GREAT INDIAN BUSTARD *Choriotis nigriceps*	37(5)
355.	HOUBARA *Chlamydotis undulata*	37(4)
356.	BENGAL FLORICAN *Eupodotis bengalensis*	37(3)
357.	LIKH or LESSER FLORICAN *Sypheotides indica*	37(6)

Family JAÇANIDAE: JAÇANAS

Plumage bronze, black, brown, and brown-and-white in combinations. Bill slender, longish, straight and compressed. Legs long, tibiae bare; toes and claws enormously elongated enabling the birds to trip lightly over floating leaves and vegetation. Wings broad with sharp metacarpal spur. Tail short except in *Hydrophasianus*, where it is narrow, long and arched. Flight feeble, with the large feet dangling behind, seldom more than 2 or 3 metres above the surface. Food, chiefly vegetable matter; also insects, molluscs, etc. Sexes alike but female larger, and polyandrous. Nest, a flimsy pad or raft of grass or weed-stems on floating vegetation. Eggs, normally 4, often laid directly on floating leaves. Incubation by the male alone. Young, nidifugous and downy.

		Plate
358.	PHEASANT-TAILED JACANA *Hydrophasianus chirurgus*	36(16)
359.	BRONZEWINGED JACANA *Metopidius indicus*	36(18)

Family HAEMATOPODIDAE: Oystercatchers

Shore birds. Bill long, compressed, slightly truncated at end, and red in colour. Tarsus short, stout and reticulated; hallux absent; anterior toes short, stout and slightly webbed. Wings long and pointed. Flight strong. Sexes alike. Feeds largely on marine molluscs. Nest, usually a depression in the ground. Eggs, 3-4.

		Plate
360.	OYSTERCATCHER *Haematopus ostralegus*	39(5), 42(1)

Family ROSTRATULIDAE: Painted Snipe

Largely crepuscular, brightly coloured terrestrial birds of reedy marshes. Plumage brown, grey, bronzy, black and white, cryptically patterned. Neck short. Bill long, slightly curved and swollen at tip. Toes long.

Wings broad. Tail short. Flight slow; strong and swift runners; effective swimmers. Food, molluscs, crustaceans, insects, worms and weed-seeds, paddy grains, etc., obtained by probing into squelchy mud or by scything movements of the bill in shallow water. Female larger and more brightly coloured than male, polyandrous, pugnacious and the dominant sex partner. Nest, a pad of grass or rushes with a slight depression in the centre, built on the ground or floating in the water. Eggs, normally 4. Incubation and raising of the chicks done by male.

		Plate
429.	PAINTED SNIPE *Rostratula benghalensis*	38(8), 41(16)

Family RECURVIROSTRIDAE: Stilts, Avocets

Waders or shore birds with white, black or brownish grey plumage. Bill long, slender and straight or upcurved. Legs moderate to extremely long and slender; feet webbed; hallux vestigial or absent. Wings long and pointed; tail short and square. Food, molluscs, worms, crustaceans, insects, and seeds of sedges and marsh plants. Nest, a hollow depression in the ground, or a raised platform of pebbles, sometimes lined with vegetable scum off the water, grass, etc. Eggs, normally 4. Incubation by both sexes. Young, nidifugous and downy.

		Plate
430.	BLACKWINGED STILT *Himantopus himantopus*	39(8), 42(8)
432.	AVOCET *Recurvirostra avosetta*	39(6), 42(10)

Family IBIDORHYNCHIDAE: Ibisbill

Greyish brown and white sandpiper-like wading birds. Bill red in colour, long, hard, slender and down-curved. Tarsus comparatively short, reticulated throughout. No hind toe; outer and middle toes connected by a deeply indented small web; web between middle and inner toes obsolete. Wings square. Sexes alike. Food, insects, molluscs, crustaceans, worms, etc. obtained by wading breast deep, ducking head and neck under water, and thrusting the long bill under the submerged pebbles. Nest, an unlined depression on a shingly islet amidst a glacier stream. Eggs 4. Incubation by both sexes.

		Plate
433.	IBISBILL *Ibidorhyncha struthersii*	39(16), 42(3)

Family DROMADIDAE: Crab Plovers

Largely crepuscular, maritime waders. Plumage mainly black and white. Bill strong and laterally compressed. Legs long and bare; middle claw pectinate. Wings long and pointed; tail short. Sexes alike. Food, chiefly crabs. nest, hole in sandbanks on coral reefs; colonial. Egg, a singleton, inordinately large.

		Plate
434.	CRAB PLOVER *Dromas ardeola*	39(7), 42(9)

Family BURHINIDAE: Stone Curlews, Thick-knees

Terrestrial cursorial birds. Plumage brown, buff and white cryptically patterned, and with conspicuous head and wing markings. Head large and broad; very large 'goggle' eyes. Bill stout. Legs long; tarsus bare, tibiotarsal ('knee') joint thickened; feet partially webbed; toes three. Wings long and pointed. Sexes alike. Food, chiefly animal matter - insects, lizards, etc. Nest, a shallow, unlined scrape on the ground. Eggs, 1-3. Incubation by both sexes, chiefly by the female.

		Plate
436.	STONE CURLEW *Burhinus oedicnemus*	39(14), 42(11)
437.	GREAT STONE PLOVER *Esacus magnirostris*	39(15), 42(12)

Family GLAREOLIDAE: Coursers, Pratincoles

Brown, grey and white cursorial (running) birds boldly marked with black, white and chestnut. Coursers: Wings short and broad; tail short; bill longish and tapering; legs long and bare; three anterior toes. Most feeding movement on the ground by running swiftly in spurts, but also capable of fast, sustained flight. Pratincoles: Wings long, narrow and pointed; tail forked; bill and legs short; hallux present. Middle toe elongate with pectinate claw. Feed on the wing on flying insects. Sexes alike or nearly so. Nest, on the ground. Eggs, 2-3.

		Plate
439.	CREAMCOLOURED COURSER *Cursorius cursor*	38(1), 41(17)
440.	INDIAN COURSER *Cursorius coromandelicus*	38(2), 41(18)
441.	JERDON'S or DOUBLEBANDED COURSER *Cursorius bitorquatus*	38(3)
442.	COLLARED PRATINCOLE *Glareola pratincola pratincola*	44(2)
443.	LARGE INDIAN PRATINCOLE *Glareola pratincola*	38(4), 44(3)
444.	SMALL INDIAN PRATINCOLE *Glareola lactea*	38(5), 44(4)

Family CHARADRIIDAE:
Plovers, Sandpipers, Snipe

Wading birds of small to medium size. Bill short and pigeon-like to long, slender and straight or downcurved. Legs short to long, with tibiae partly bare in many species. Wings long and pointed; tail short to medium. Flight strong, swift and well sustained. Sexes may be nearly alike, or female may be much smaller and/or duller coloured than male. Food, small invertebrates, molluscs, insects, worms and some vegetable matter obtained by probing in soft mud. Nest, on the ground. Eggs, 2-5.

		Plate
362.	WHITETAILED LAPWING *Vanellus leucurus*	38(14), 39(4)
363.	SOCIABLE LAPWING *Vanellus gregarius*	38(15)
364.	LAPWING *Vanellus vanellus*	38(9), 39(1)
365.	GREYHEADED LAPWING *Vanellus cinereus*	38(12), 39(3)
366.	REDWATTLED LAPWING *Vanellus indicus*	38(11), 39(2)
369.	SPURWINGED LAPWING *Vanellus spinosus*	38(10)
370.	YELLOW-WATTLED LAPWING *Vanellus malabaricus*	38(13)
371.	GREY or BLACKBELLIED PLOVER *Pluvialis squatarola*	38(6), 44(1)
372.	GOLDEN PLOVER *Pluvialis apricaria*	44(5)
373.	EASTERN GOLDEN PLOVER *Pluvialis dominica*	38(7), 44(7)
374.	LARGE SAND PLOVER *Charadrius leschenaultii*	40(6)

		Plate
376.	CASPIAN SAND PLOVER *Charadrius asiaticus*	40(7)
378.	RINGED PLOVER *Charadrius hiaticula*	40(2), 41(1)
380.	LITTLE RINGED PLOVER *Charadrius dubius*	40(3), 41(2)
381.	KENTISH PLOVER *Charadrius alexandrinus*	40(4), 41(3)
383.	LONGBILLED RINGED PLOVER *Charadrius placidus*	40(1)
384.	LESSER SAND PLOVER *Charadrius mongolus*	40(5), 41(4)
385.	WHIMBREL *Numenius phaeopus*	39(9), 42(7)
388.	CURLEW *Numenius arquata*	39(13), 42(4)
389.	BLACKTAILED GODWIT *Limosa limosa*	39(10), 42(5)
391.	BARTAILED GODWIT *Limosa lapponica*	39(11), 42(6)
392.	SPOTTED or DUSKY REDSHANK *Tringa erythropus*	44(9)
393.	REDSHANK *Tringa totanus*	43(1), 44(12)
395.	MARSH SANDPIPER *Tringa stagnatilis*	43(2)
396.	GREENSHANK *Tringa nebularia*	43(3), 44(10)
397.	GREEN SANDPIPER *Tringa ochropus*	43(6), 44(8)
398.	WOOD SANDPIPER *Tringa glareola*	43(5), 44(11)
399.	SPOTTED GREENSHANK *Tringa guttifer*	43(4)
400.	TEREK SANDPIPER *Tringa terek*	43(7), 44(14)
401.	COMMON SANDPIPER *Tringa hypoleucos*	43(9), 44(13)
402.	TURNSTONE *Arenaria interpres*	43(8), 44(6)
403.	SNIPEBILLED GODWIT or ASIAN DOWITCHER *Limnodromus semipalmatus*	39(12), 42(2)
404.	SOLITARY SNIPE *Gallinago solitaria*	106(6)
405.	WOOD SNIPE *Gallinago nemoricola*	106(4)
406.	PINTAIL SNIPE *Gallinago stenura*	41(15), 106(3)
407.	SWINHOE'S SNIPE *Gallinago megala*	106(2)
408.	GREAT SNIPE *Gallinago media*	106(1)
409.	FANTAIL SNIPE *Gallinago gallinago*	41(13), 43(11), 106(5)
410.	JACK SNIPE *Gallinago minima*	43(10), 106(7)
411.	WOODCOCK *Scolopax rusticola*	41(14), 43(13)
412.	KNOT *Calidris canutus*	40(13), 41(13)
413.	EASTERN KNOT *Calidris tenuirostris*	40(12), 41(11)
414.	SANDERLING *Calidris alba*	40(19), 41(5)
416.	LITTLE STINT *Calidris minuta*	40(8), 41(8)
417.	TEMMINCK'S STINT *Calidris temminckii*	40(9), 41(9)
418.	LONGTOED STINT *Calidris subminuta*	40(10)
419.	ASIAN PECTORAL SANDPIPER *Calidris acuminata*	40(11)
420.	DUNLIN *Calidris alpina*	40(14), 41(6)
422.	CURLEW-SANDPIPER *Calidris testacea*	40(15), 41(7)
423.	SPOONBILLED SANDPIPER *Eurynorhynchus pygmeus*	40(18)
424.	BROADBILLED SANDPIPER *Limicola falcinellus*	40(16)
426.	RUFF and REEVE *Philomachus pugnax*	43(12), 44(15)
428.	REDNECKED PHALAROPE *Phalaropus lobatus*	40(17), 41(12)

Family STERCORARIIDAE: Skuas, Jaegers

Pelagic, gull-like birds, many with dark and light plumage phases. Body stout. Bill strong, rounded and strongly hooked, with a horny cere. Legs short; feet stout and fully webbed; claws small, but curved and sharp. Wings long and pointed; tail wedge-shaped. Flight powerful and swift. Sexes alike. Skuas are predatory on the eggs and chicks of terns, penguins, and other colonial nesting sea birds. They also attack and chase other birds, forcing them to either drop their lawful prize or disgorge their crop contents, which they then retrieve in mid-air. They are birds of the Arctic and Antarctic zones. Within our limits, the records are mainly of accidental waifs storm-tossed during heavy monsoon gales.

		Plate
445.	ANTARCTIC SKUA *Catharacta skua antarctica*	45(1)
446a.	ANTARCTIC SKUA (MACCORMICK'S SKUA) *Catharacta skua maccormicki*	45(4)
447.	POMATORHINE SKUA or JAEGER *Stercorarius pomarinus*	45(2)
448.	PARASITIC or RICHARDSON'S SKUA or JAEGER *Stercorarius parasiticus*	45(5)

Family LARIDAE: Gulls, Terns

Gulls: Gregarious, heavy-bodied aquatic birds. Plumage chiefly a combination of grey, white and black. Bill slender to heavy, sharply pointed or blunt and slightly hooked. Legs short; feet webbed; hallux small or vestigial. Wings long and pointed; tail square or forked. Sexes alike. Flight strong. Feed by catching fish, scavenging on various animal and vegetable matter, and by piracy like skuas. Roost and breed in colonies, often of great size. Nest, on the ground, cliffs, in trees, or floating. May be a bulky structure of grass, etc. or a skimpily lined or unlined scrape. Eggs, 2-4. Incubation by both sexes. Terns are more lightly built than gulls, with longer, narrower wings and a different style of flight. They rest and roost on rocks or mudbanks, and in spite of their webbed feet rarely settle on water. They capture living prey (fish, crabs etc.) by scooping it up from the surface in flight or diving vertically from the air and going under water momentarily. Nesting colonial.

		Plate
449.	SOOTY GULL *Larus hemprichii*	45(7)
450.	HERRING GULL *Larus argentatus*	45(8)
452.	LESSER BLACKBACKED GULL *Larus fuscus*	45(9)
453.	GREAT BLACKHEADED GULL *Larus ichthyaetus*	45(6)
454.	BROWNHEADED GULL *Larus brunnicephalus*	45(13)
455.	BLACKHEADED GULL *Larus ridibundus*	45(11)
456.	SLENDERBILLED GULL *Larus genei*	45(12)
457.	LITTLE GULL *Larus minutus*	45(10)
458.	WHISKERED TERN *Chlidonias hybrida*	46(3)
459.	WHITEWINGED BLACK TERN *Chlidonias leucopterus*	46(2)
459a.	BLACK TERN *Chlidonias niger*	46(1)
460.	GULLBILLED TERN *Gelochelidon nilotica*	46(15)
462.	CASPIAN TERN *Hydroprogne caspia*	46(10)
463.	INDIAN RIVER TERN *Sterna aurantia*	46(9)
464.	COMMON TERN *Sterna hirundo*	46(8)
466.	ROSEATE or ROSY TERN *Sterna dougallii*	46(6)

		Plate
467.	WHITECHEEKED TERN *Sterna repressa*	46(7)
468.	BLACKNAPED TERN *Sterna sumatrana*	46(12)
470.	BLACKBELLIED TERN *Sterna acuticauda*	46(13)
471.	BROWNWINGED TERN *Sterna anaethetus*	46(19)
474.	SOOTY TERN *Sterna fuscata*	46(16)
475.	LITTLE TERN *Sterna albifrons*	46(4)
478.	LARGE CRESTED TERN *Sterna bergii*	46(11)
479.	INDIAN LESSER CRESTED TERN *Sterna bengalensis*	46(14)
480.	SANDWICH TERN *Sterna sandvicensis*	46(18)
481.	NODDY TERN *Anous stolidus*	46(17)
483.	INDIAN OCEAN WHITE TERN, FAIRY TERN *Gygis alba*	46(5)
484.	INDIAN SKIMMER *Rynchops albicollis*	45(3)

Family PTEROCLIDIDAE: Sandgrouse

Terrestrial pigeon-like birds. Plumage chiefly sandy brown patterned with black spots and vermiculations, affording perfect camouflage to the birds in their native semi-desert environment. They have an exceptionally powerful flight, and in spite of their short legs are also good walkers and runners. Bill small but stout. Wings long and pointed; tail moderately long and pointed. Legs and toes short; claws short and thick. Sexes differ. Sandgrouse have a typical habit of flying in flocks to water regularly, a few hours after sunrise and during early dusk; this flight often covers long distances. Food, chiefly vegetarian — seeds and shoots of desert grasses and weeds. Nest, a shallow unlined scrape in the ground. Eggs, 2-3. Incubation by both sexes. Young precocial.

		Plate
485.	TIBETAN SANDGROUSE *Syrrhaptes tibetanus*	47(9)
485a.	PALLAS'S SANDGROUSE *Syrrhaptes paradoxus*	47(8)
486.	LARGE PINTAIL SANDGROUSE *Pterocles alchata*	47(7)
487.	INDIAN SANDGROUSE *Pterocles exustus*	47(4)
488.	SPOTTED SANDGROUSE *Pterocles senegallus*	47(6)
489.	IMPERIAL SANDGROUSE *Pterocles orientalis*	47(1)
490.	CORONETTED SANDGROUSE *Pterocles coronatus*	47(2)
491.	CLOSEBARRED SANDGROUSE *Pterocles indicus arabicus*	47(3)
492.	PAINTED SANDGROUSE *Pterocles indicus indicus*	47(5)

Family COLUMBIDAE: Pigeons, Doves

Arboreal or terrestrial birds. Plumage dense and soft. Body compact. Head small; neck short. Bill of medium length and slender to stout, with a naked cere. Legs short to fairly long. Flight swift and powerful. Sexes alike in most species. Food chiefly seeds, grain, drupes, etc. Drink water by immersing the bill and sucking continuously. Nest, a sketchy platform of a few sticks in trees, on ledges or in holes in cliffs etc. Eggs, normally 2. Incubation and care of young by both sexes. Initially the nestlings are fed on a secretion of the parent's crop known as 'pigeon's milk'. Voice, cooing or booming calls or mellow musical whistles. Many species migratory. The homing instinct of some domestic breeds extensively exploited for carrying messages prior to the advent of wireless telegraphy and even during the two recent World Wars.

		Plate
493.	PINTAILED GREEN PIGEON *Treron apicauda*	49(5)
494.	WEDGETAILED GREEN PIGEON *Treron sphenura*	49(1)
495.	THICKBILLED GREEN PIGEON *Treron curvirostra*	49(4)
496.	POMPADOUR or GREYFRONTED GREEN PIGEON *Treron pompadora*	49(3)
501.	ORANGEBREASTED GREEN PIGEON *Treron bicincta*	49(6)
503.	YELLOWLEGGED GREEN PIGEON *Treron phoenicoptera*	49(2)
506.	GREEN IMPERIAL PIGEON *Ducula aenea*	48(9)
509.	PIED IMPERIAL PIGEON *Ducula bicolor*	48(10)
510.	IMPERIAL PIGEON *Ducula badia*	48(15)
513.	SNOW PIGEON *Columba leuconota*	48(5)
515.	HILL PIGEON *Columba rupestris*	48(4)
516.	BLUE ROCK PIGEON *Columba livia*	48(1)
518.	EASTERN STOCK PIGEON *Columba eversmanni*	48(3)
519.	WOOD PIGEON *Columba palumbus*	48(6)
520.	SPECKLED WOOD PIGEON *Columba hodgsonii*	48(14)
521.	NILGIRI WOOD PIGEON *Columba elphinstonii*	48(7)
522.	CEYLON WOOD PIGEON *Columba torringtoni*	48(12)
523.	ASHY WOOD PIGEON *Columba pulchricollis*	48(8)
524.	PURPLE WOOD PIGEON *Columba punicea*	48(11)
525.	ANDAMAN WOOD PIGEON *Columba palumboides*	48(13)
526.	BARTAILED CUCKOO-DOVE *Macropygia unchall*	49(8)
527.	ANDAMAN CUCKOO-DOVE *Macropygia rufipennis*	49(10)
529.	TURTLE DOVE *Streptopelia turtur*	49(15)
530.	RUFOUS TURTLE DOVE *Streptopelia orientalis*	49(16)
534.	INDIAN RING DOVE *Streptopelia decaocto*	49(12)
535.	RED TURTLE DOVE *Streptopelia tranquebarica*	49(11)
537.	SPOTTED DOVE *Streptopelia chinensis*	49(14)
541.	LITTLE BROWN or SENEGAL DOVE *Streptopelia senegalensis*	49(13)
542.	EMERALD or BRONZEWINGED DOVE *Chalcophaps indica*	49(9)
544b.	NICOBAR PIGEON *Caloenas nicobarica*	49(7)
EL.	STOCK PIGEON *Columba oenas*	48(2)

Family PSITTACIDAE: Parrots

Brightly coloured arboreal fruit- and grain-eating birds. All Indian species chiefly green. They include some of the most abundant and destructive bird pests of agriculture and horticulture in India. Plumage sparse. Bill short, stout, strongly hooked; upper mandible loosely articulated with the skull and capable of kinetic movement. Tongue thick and fleshy. Feet zygodactylous. Wings rounded. Flight swift. Sexes alike or differing slightly. Food, purely vegetarian — fruit, berries, nuts, grain, etc. Voice loud, high-pitched and harsh. Many species can be trained to imitate a few words of human speech in captivity. Nest, usually in holes in trees. Eggs 2-5.

		Plate
546.	ALEXANDRINE PARAKEET *Psittacula eupatria*	50(3)
550.	ROSERINGED PARAKEET *Psittacula krameri*	50(9)

		Plate
551.	REDBREASTED PARAKEET *Psittacula alexandri*	50(11)
553.	NICOBAR PARAKEET *Psittacula caniceps*	50(1)
554.	LORD DERBY'S PARAKEET *Psittacula derbyana*	50(2)
555.	REDCHEEKED PARAKEET *Psittacula longicauda*	50(8)
558.	BLOSSOMHEADED PARAKEET *Psittacula cyanocephala*	50(5)
559.	EASTERN BLOSSOMHEADED PARAKEET *Psittacula roseata*	50(7)
562.	SLATYHEADED PARAKEET *Psittacula himalayana*	50(6)
563.	EASTERN SLATYHEADED PARAKEET *Psittacula finschii*	50(4)
564.	BLUEWINGED PARAKEET *Psittacula columboides*	50(10)
565.	LAYARD'S PARAKEET *Psittacula calthropae*	50(12)
566.	INDIAN LORIKEET *Loriculus vernalis*	50(14)
568.	CEYLON LORIKEET *Loriculus beryllinus*	50(13)

Family CUCULIDAE: Cuckoos

Arboreal or terrestrial. Bill curved, stout, and sometimes heavy. Legs short; feet zygodactyl. Wings medium to long; tail medium to extremely long, often graduated. Sexes usually alike. Food, insects, molluscs, lizards, snakes and other small vertebrates; also fruits. Voice, loud notes — mostly harsh and shrieking but some quite melodious — monotonously repeated. This family is well known for its habit of brood-parasitism, building no nests, but laying the eggs in the nests of other birds, and foisting the labour of incubating the eggs and rearing the young upon the dupes. Some species are non-parasitic, having normal habits as in other birds. Young nidicolous.

		Plate
569.	REDWINGED CRESTED CUCKOO *Clamator coromandus*	52(3)
571.	PIED CRESTED CUCKOO *Clamator jacobinus*	52(2)
572.	LARGE HAWK-CUCKOO *Cuculus sparverioides*	51(1)
573.	COMMON HAWK-CUCKOO or BRAINFEVER BIRD *Cuculus varius*	51(3)
575.	HODGSON'S HAWK-CUCKOO *Cuculus fugax*	51(2)
576.	INDIAN CUCKOO *Cuculus micropterus*	51(10)
578.	THE CUCKOO *Cuculus canorus*	51(11)
581.	SMALL CUCKOO *Cuculus poliocephalus*	51(9)
582.	INDIAN BANDED BAY CUCKOO *Cacomantis sonneratii*	51(6)
584.	INDIAN PLAINTIVE CUCKOO *Cacomantis passerinus*	51(8)
585.	RUFOUSBELLIED PLAINTIVE CUCKOO *Cacomantis merulinus*	51(5)
586.	EMERALD CUCKOO *Chalcites maculatus*	51(4)
587.	VIOLET CUCKOO *Chalcites xanthorhynchus*	51(7)
588.	DRONGO-CUCKOO *Surniculus lugubris*	69(11)
590.	KOEL *Eudynamys scolopacea*	71(1)
593.	LARGE GREENBILLED MALKOHA *Rhopodytes tristis*	52(4)
595.	SMALL GREENBILLED MALKOHA *Rhopodytes viridirostris*	52(1)
598.	SIRKEER CUCKOO *Taccocua leschenaultii*	52(5)
599.	REDFACED MALKOHA *Phaenicophaeus pyrrhocephalus*	52(7)
600.	CROW-PHEASANT or COUCAL *Centropus sinensis*	52(8)
603.	ANDAMAN CROW-PHEASANT *Centropus (sinensis) andamanensis*	52(10)

		Plate
604.	CEYLON COUCAL *Centropus chlororhynchus*	52(9)
605.	LESSER COUCAL *Centropus toulou*	52(6)

Family STRIGIDAE: Owls

Arboreal, chiefly nocturnal birds of prey. Plumage long and soft; grey, brown, chestnut, black or white, and barred, streaked or vermiculated. Head large; eyes very large, and forwardly directed. A facial disc and 'horns' present in many species. Bill short, strong and hooked. Flank feathers greatly elongated; tarsus and toes feathered; outer toe reversible; claws strongly hooked. Wings broad and rounded. Sexes generally alike, but female larger. Flight buoyant and, in most species, silent. Food, entirely animal, principally rodents; also other small creatures. Nest, in holes or hollows. Young nidicolous and downy. Care of young by both sexes.

		Plate
606.	BARN OWL *Tyto alba*	54(16)
608.	GRASS OWL *Tyto capensis*	54(17)
609.	BAY OWL *Phodilus badius*	54(15)
612.	SPOTTED SCOPS OWL *Otus spilocephalus*	54(2)
614.	STRIATED or PALLID SCOPS OWL *Otus brucei*	54(1)
617.	SCOPS OWL *Otus scops*	54(3)
623.	COLLARED SCOPS OWL *Otus bakkamoena*	54(4)
627.	EAGLE-OWL or GREAT HORNED OWL *Bubo bubo*	53(1)
628.	FOREST EAGLE-OWL *Bubo nipalensis*	53(2)
630.	DUSKY HORNED OWL *Bubo coromandus*	53(3)
631.	BROWN FISH OWL *Bubo zeylonensis*	53(8)
633.	TAWNY FISH OWL *Bubo flavipes*	53(11)
635.	COLLARED PIGMY OWLET *Glaucidium brodiei*	54 (7)
636.	JUNGLE OWLET *Glaucidium radiatum*	54(8)
638.	Ssp. *castononotum* of 636	54(5)
639.	BARRED OWLET *Glaucidium cuculoides*	54(6)
642.	BROWN HAWK-OWL *Ninox scutulata*	54(14)
645.	Ssp. *obscura* of 642	54(13)
648, 649.	LITTLE OWL *Athene noctua*	54(10)
652.	SPOTTED OWLET *Athene brama*	54(12)
653.	FOREST SPOTTED OWLET *Athene blewitti*	54(11)
654.	HUME'S WOOD OWL *Strix butleri*	53(10)
657.	MOTTLED WOOD OWL *Strix ocellata*	53(5)
659.	BROWN WOOD OWL *Strix leptogrammica*	53(4)
662.	HIMALAYAN WOOD OWL *Strix aluco*	53(9)
663.	LONGEARED OWL *Asio otus*	53(6)
664.	SHORTEARED OWL *Asio flammeus*	53(7)
665.	TENGMALM'S or BOREAL OWL *Aegolius funereus*	54(9)

Family PODARGIDAE: Frogmouths

Crepuscular and nocturnal birds with soft, silky plumage — cryptically patterned in brown, tawny, grey and black. Neck short and thick. Bill broad, flat and triangular, very wide at gape and hooked at tip; the base overhung by bristly feathers. Legs short; feet small and weak; middle toe elongated. Wings rounded. Sexes dimorphic. Spend the day perched across a bare branch, body tilted at an upward angle bill pointing to sky, simulating a broken stump. Food, insects. Nest, a small pad of leaves, twigs, etc., lined with down from the bird's underplumage; placed in the fork of, or on a horizontal branch. Eggs, 1-3. Incubation by both sexes.

		Plate
666.	CEYLON FROGMOUTH *Batrachostomus moniliger*	55(1)
667.	HODGSON'S FROGMOUTH *Batrachostomus hodgsoni*	55(2)

Family CAPRIMULGIDAE: Nightjars, Goatsuckers

Crepuscular or nocturnal soft plumaged birds, cryptically patterned in rufous, buff, grey, black and white. Head and eyes large; neck shortish. Bill weak and small; gape wide; middle toe long with a pectinate claw. Wings long and pointed. Sexes nearly alike. Food, inseots, captured in flight. Eggs, 1-2, laid on the bare ground.

		Plate
669.	GREAT EARED NIGHTJAR *Eurostopodus macrotis*	55(10)
671.	INDIAN JUNGLE NIGHTJAR *Caprimulgus indicus*	55(6)
673.	EUROPEAN NIGHTJAR *Caprimulgus europaeus*	55(4)
673a.	EGYPTIAN NIGHTJAR *Caprimulgus aegyptius*	55(5)
674.	SYKES'S NIGHTJAR *Caprimulgus mahrattensis*	55(3)
675.	LONGTAILED NIGHTJAR *Caprimulgus macrurus*	55(9)
680.	COMMON INDIAN NIGHTJAR *Caprimulgus asiaticus*	55(8)
682.	FRANKLIN'S or ALLIED NIGHTJAR *Caprimulgus affinis*	55(7)

Family APODIDAE: Swifts

The most aerial of all birds. Body compact; neck short. Plumage chiefly brown or blackish, with patches of white or grey. Bill small, slightly decurved, with a very large gape. Legs short; feet small with hind toe completely reversible and needle-sharp curved claws, useless for walking or perching. Wings bow-shaped — long, narrow and pointed. Sexes alike. Swifts are capable of sustained high-speed flight, often spending all day on the wing; they rest by clinging to rough, vertical surfaces. Food, insects hawked in flight. Nest, of grass, tow, etc., glued together with saliva and attached to a vertical surface. Some species build their nests entirely of saliva; these nests are edible and of commercial value. Eggs, 1-6. Incubation by both sexes.

		Plate
683.	HIMALAYAN SWIFTLET *Collocalia brevirostris.*	56(2)
685.	INDIAN EDIBLE-NEST SWIFTLET *Collocalia unicolor*	56(8)
687.	WHITEBELLIED SWIFTLET *Collocalia esculenta*	56(3)
688.	WHITETHROATED SPINETAIL SWIFT *Chaetura caudacuta*	56(14)
690.	COCHINCHINA SPINETAIL SWIFT *Chaetura cochinchinensis*	56(11)
691.	LARGE BROWNTHROATED SPINETAIL SWIFT *Chaetura gigantea*	56(10)
692.	WHITERUMPED SPINETAIL *Chaetura sylvatica*	56(1)
693.	ALPINE SWIFT *Apus melba*	56(12)

		Plate
696.	THE SWIFT *Apus apus*	56(6)
697.	PALLID SWIFT *Apus pallidus*	56(5)
698.	DARKBACKED SWIFT *Apus acuticauda*	56(9)
699.	LARGE WHITERUMPED SWIFT *Apus pacificus*	56(7)
703.	HOUSE SWIFT *Apus affinis*	56(4)
707.	PALM SWIFT *Cypsiurus parvus*	56(13)
709.	CRESTED TREE SWIFT *Hemiprocne longipennis*	5(1)

Family TROGONIDAE: Trogons

Brightly coloured arboreal birds with soft, dense plumage, and a thin delicate skin. Neck short. Eyes large; brightly coloured bare orbital rings. Bill short and usually brightly coloured. Legs and feet small and weak, the first and second toes turned back. Wings short and rounded; tail long and graduated. Sexes dimorphic. Usually solitary, may be in pairs or in small groups. Perch erect and motionless on branches for long periods, statue-like. Flight swift, but generally not over long distances. Food, insects and berries. Nest, in holes in trees. Eggs, 2-4. Incubation by both sexes. Young, altricial and nidicolous.

		Plate
712.	MALABAR TROGON *Harpactes fasciatus*	57(3)
715.	REDHEADED TROGON *Harpactes erythrocephalus*	57(1)
716.	WARD'S TROGON *Harpactes wardi*	57(2)

Family ALCEDINIDAE: Kingfishers

Usually of blue, green purple, brown or black-and-white plumage. Body compact; neck short; bill massive, long, straight and pointed. Legs short; feet syndactyl. Wings short and rounded. Sexes generally alike. Flight direct and swift. Non-social birds, adapted for life chiefly at inland waters; though some species are largely maritime and others largely terrestrial. The 'aquatic' forms feed on fish obtained by diving headlong into the water; the others live also on large insects and small vertebrates. Nest, burrows in river banks or holes in trees. Eggs, 2-7. Incubation by both sexes.

		Plate
717.	HIMALAYAN PIED KINGFISHER *Ceryle lugubris*	58(1)
719.	LESSER PIED KINGFISHER *Ceryle rudis*	58(3)
721.	BLYTH'S or GREAT BLUE KINGFISHER *Alcedo hercules*	58(7)
722.	COMMON KINGFISHER *Alcedo atthis*	58(6)
725.	BLUE-EARED KINGFIHER *Alcedo meninting*	58(5)
727.	THREETOED KINGFISHER *Ceyx erithacus*	58(4)
727a.	THREETOED KINGFISHER *Ceyx erithacus rufidorsus*	58(2)
729.	BROWNWINGED KINGFISHER *Pelargopsis amauroptera*	58(9)
730.	STORKBILLED KINGFISHER *Pelargopsis capensis*	58(12)
733.	RUDDY KINGFISHER *Halcyon coromanda*	58(11)
735.	WHITEBREASTED KINGFISHER *Halcyon smyrnensis*	58(13)
739.	BLACKCAPPED KINGFISHER *Halcyon pileata*	58(8)
740.	WHITECOLLARED KINGFISHER *Halcyon chloris*	58(10)

Family MEROPIDAE: Bee-eaters

Gregarious sleek-looking birds with soft and compact plumage. Most species largely green, and generally with a black stripe from base of bill through eye. Bill long, slender, laterally compressed and decurved. Legs small, feet and toes slender. Wings long and pointed; tail longish, with the central pair of rectrices usually elongated as pins. Sexes nearly alike. Food, insects, chiefly bees and wasps caught in flight. Nest, self-excavated holes in earth banks, or sloping tunnels in the ground. Eggs, 2-8. Incubation and nest-feeding by both sexes.

		Plate
744.	CHESTNUTHEADED BEE-EATER *Merops leschenaulti*	57(4)
746.	EUROPEAN BEE-EATER *Merops apiaster*	57(7)
747.	BLUECHEEKED BEE-EATER *Merops superciliosus*	57(6)
748.	BLUETAILED BEE-EATER *Merops philippinus*	57(9)
750.	GREEN BEE-EATER *Merops orientalis*	57(10)
753.	BLUEBEARDED BEE-EATER *Nyctyornis athertoni*	57(5)

Family CORACIIDAE: Rollers

Crow-sized arboreal birds chiefly of striking brown and brilliant blue plumage. Head large; neck short. Bill wide, strong and slightly hooked. Legs short; feet strong; second and third toes basally united. Wings long; tail longish. Sexes nearly alike. Flight strong, often accompanied, especially during courtship display, by harsh cries and a series of bizarre aerobatics — nosediving, rolling, looping-the-loop, etc. Food, chiefly insects caught in air; also small crawling animals. Nest, unlined tree holes. Eggs, 2-4. Incubation by both sexes. Rollers are noisy birds, and often aggressive and quarrelsome.

		Plate
754.	EUROPEAN ROLLER *Coracias garrulus*	57(8)
755.	INDIAN ROLLER *Coracias benghalensis*	57(11)
759.	BROADBILLED ROLLER *Eurystomus orientalis*	57(12)

Family UPUPIDAE: Hoopoes

Pinkish cinnamon to chestnut coloured birds with black, white and buff zebra-like stripes on back and wings, and a long, conspicuous black-tipped fan-like crest. Bill long and slender. Tarsi short, toes long. Wings broad and rounded; tail square. Sexes nearly alike. Flight undulating. Food, largely insects probed out of the soil, using the bill as forceps. Song, a deep mellow *hoo-po* or *hoo-po-po* repeated at short intervals. Nest, holes in trees or walls; notoriously insanitary and evil-smelling from the bird's unremoved droppings and food remains. Eggs, 5-7. Incubation by female, who is fed by her mate throughout this period. Nestlings fed by both parents; when threatened, the nestlings squirt their foul-smelling liquid faeces at the intruder.

		Plate
763.	HOOPOE *Upupa epops*	58(14)

Family BUCEROTIDAE: Hornbills

Large, chiefly arboreal birds with wiry plumage, white, black, grey or brown in combinations. Bill enormous, brightly coloured, curved, and usually with a large casque on culmen. Bare brightly coloured skin around eyes; eyelashes bristly and prominent. Legs short; feet broad-soled and syndactyl. Wings broad; tail long. Food, fruit, insects, small animals, etc. Nest, holes in mature forest trees. The female walls herself into the nest, leaving a small aperture through which the male feeds her during the incubation period, and in many species till the young fledge out. The female is believed to undergo an accelerated wing moult during her incarceration. Eggs, 1-6.

		Plate
767.	COMMON GREY HORNBILL *Tockus birostris*	59(2)
768.	MALABAR GREY HORNBILL *Tockus griseus*	59(5)
770.	WHITETHROATED BROWN HORNBILL *Ptilolaemus tickelli*	59(3)
771.	RUFOUSNECKED HORNBILL *Aceros nipalensis*	59(9)
772.	WREATHED HORNBILL *Rhyticeros undulatus*	59(7)
773.	NARCONDAM HORNBILL *Rhyticeros plicatus (narcondami)*	59(1)
774.	INDIAN PIED HORNBILL *Anthracoceros malabaricus*	59(6)
775.	MALABAR PIED HORNBILL *Anthracoceros coronatus*	59(8)
776.	GREAT PIED HORNBILL *Buceros bicornis*	59(4)

Family CAPITONIDAE: Barbets

Small, stocky arboreal birds with large head. Plumage coarse and sparse, generally brightly coloured. Feather tufts present over the nostrils or well-developed rictal and chin bristles. Bill large, heavy, slightly curved and pointed. Legs short and strong; feet large and zygodactyl. Wings rounded; flight weak. Sexes generally alike. Food, fruit and insects. Calls, loud and far carrying, monotonously repeated. Nest, self-excavated holes in tree trunks and branches. Eggs, 2-4. Incubation by both sexes.

		Plate
778.	GREAT HILL BARBET *Megalaima virens*	60(4)
782.	LARGE GREEN BARBET *Megalaima zeylanica*	60(3)
784.	LINEATED BARBET *Megalaima lineata*	60(1)
785.	SMALL GREEN BARBET *Megalaima viridis*	60(2)
786.	YELLOWFRONTED BARBET *Megalaima flavifrons*	60(7)
787.	GOLDENTHROATED BARBET *Megalaima franklinii*	60(6)
788.	BLUETHROATED BARBET *Megalaima asiatica*	60(5)
789.	BLUE-EARED BARBET *Megalaima australis*	60(11)
790.	CRIMSONTHROATED BARBET *Megalaima rubricapilla malabarica*	60(9)
791.	CRIMSONTHROATED BARBET *Megalaima rubricapilla rubricapilla*	60(10)
792.	CRIMSONBREASTED BARBET or COPPERSMITH *Megalaima haemacephala*	60(8)

Family: INDICATORIDAE: Honeyguide

Small, solitary, dull-coloured arboreal birds, brown, olive and grey above and lighter below. Skin very thick and tough. Bill short, stout and blunt to slender and pointed. Tarsus short; feet zygodactyl; toes strong; claws long and hooked. Wings long and pointed tall somewhat graduated. Flight swift. Food, bees and other hymenopterans; largely also beeswax from abandoned combs. Reputed to have the remarkable habit of guiding humans and other melliphagous animals to bees' nests. Habits and breeding biology very little known. Most African species believed to be brood-parasitic.

		Plate
794.	HONEYGUIDE *Indicator xanthonotus*	104(1)

Family PICIDAE: Woodpeckers

Arboreal. Plumage black, white, yellow, red, brown or green. Head large; neck slender but very strong. Tongue extremely long, cylindrical, protrusible and barb-tipped, used for catching insects and skewering out beetle larvae from pupal galleries in the wood. Bill strong and chisel-like. Legs short; feet zygodactyl and very strong. Wings strong and rounded; tail rounded or wedge-shaped, the rectrices stiff and pointed, useful as a support when climbing a tree trunk. Sexes dimorphic. Flight strong and 'bounding'. Food, chiefly insects; also fruits. Nest, unlined holes in tree trunks and branches. Eggs, 2-6 or 8.

		Plate
796.	WRYNECK *Jynx torquilla*	62(1)
798.	SPECKLED PICULET *Picumnus innominatus*	61(2)
800.	RUFOUS PICULET *Sasia ochracea*	61(5)
804.	RUFOUS WOODPECKER *Micropternus brachyurus*	62(14)
807.	SCALYBELLIED GREEN WOODPECKER *Picus squamatus*	62(6)
808.	LITTLE SCALYBELLIED GREEN WOODPECKER *Picus myrmecophoneus*	62(7)
809.	BLACKNAPED GREEN WOODPECKER *Picus canus*	62(5)
813.	LARGE YELLOWNAPED WOODPECKER *Picus flavinucha*	62(4)
815.	SMALL YELLOWNAPED WOODPECKER *Picus chlorolophus*	62(2)
816.	Ssp. *chlorigaster* of 815	62(3)
819.	LESSER GOLDENBACKED WOODPECKER *Dinopium benghalense*	60(13)
823.	Ssp. *psarodes* of 819	60(14)
824.	HIMALAYAN GOLDENBACKED THREETOED WOODPECKER *Dinopium shorii*	60(15)
825.	INDIAN GOLDENBACKED THREETOED WOODPECKER *Dinopium javanense*	60(16)
827.	PALEHEADED WOODPECKER *Gecinulus grantia*	62(12)
828.	HIMALAYAN GREAT SLATY WOODPECKER *Mulleripicus pulverulentus*	62(10)
830.	INDIAN GREAT BLACK WOODPECKER *Dryocopus javensis*	62(8)
831.	Ssp. *hodgei* of 830	62(9)
833.	RUFOUSBELLIED WOODPECKER or SAPSUCKER *Hypopicus hyperythrus*	61(11)
834.	GREAT SPOTTED or REDCROWNED PIED WOODPECKER *Picoides major*	61(14)
835.	SIND PIED WOODPECKER *Picoides assimilis*	61(12)
836.	Ssp. *albescens* of 837	61(16)
837.	HIMALAYAN PIED WOODPECKER *Picoides himalayensis*	61(15)
838.	DARJEELING PIED WOODPECKER *Picoides darjellensis*	61(18)
840.	CRIMSONBREASTED PIED WOODPECKER *Picoides cathpharius*	61(13)

		Plate
842.	BROWNFRONTED PIED WOODPECKER *Picoides auriceps*	61(6)
844.	STRIPEBREASTED PIED WOODPECKER *Picoides atratus*	61(8)
845.	FULVOUSBREASTED PIED WOODPECKER *Picoides macei*	61(10)
847.	YELLOWFRONTED PIED WOODPECKER *Picoides mahrattensis*	61(9)
850.	GREYCROWNED PIGMY WOODPECKER *Picoides canicapillus*	61(3)
851.	PIGMY WOODPECKER *Picoides nanus*	61(1)
855.	THREETOED WOODPECKER *Picoides tridactylus*	61(7)
856.	HEARTSPOTTED WOODPECKER *Hemicircus canente*	61(4)
857.	REDEARED BAY WOODPECKER *Blythipicus pyrrhotis*	62(13)
858.	BLACKBACKED WOODPECKER *Chrysocolaptes festivus*	60(12)
861.	LARGER GOLDENBACKED WOODPECKER *Chrysocolaptes lucidus*	60(17)
863.	Ssp. *stricklandi* of 861	60(18)
EL.	WHITEMANTLED WOODPECKER *Dendrocopos leucopterus*	61(17)
EL.	LACED WOODPECKER *Picus vittatus*	62(11)

Family EURYLAIMIDAE: Broadbills

Brightly coloured arboreal birds. Body stout; head broad; eyes large. Bill broad, flattened and hooked, with a wide gape. Legs short; feet strong and syndactyl; toes and claws long. Wings rounded; tail short and square or long, slender and graduated. Sexes dimorphic. Feed largely on insects. Usually tame and confiding. Nest, a conspicuous purse-like structure suspended from a branch with no attempt at concealment. Eggs, 1-5.

		Plate
864.	COLLARED BROADBILL *Serilophus lunatus*	63(3)
865.	LONGTAILED BROADBILL *Psarisomus dalhousiae*	63(1)

Family PITTIDAE: Pittas

Plump, solitary birds, largely terrestrial and generally brightly coloured. Neck short. Bill strong and slightly curved. Legs strong and long; feet large. Wings short and rounded; tail very short. Generally hop on the ground but perch on trees to sing and roost. Flight strong. Food, insects, grubs and other invertebrates. Voice, rich whistles and trilling calls. Nest, a globular structure of twigs and grass, on the ground or in trees. Eggs, 2-7. Incubation by both sexes.

		Plate
866.	BLUENAPED PITTA *Pitta nipalensis*	63(10)
867.	INDIAN PITTA *Pitta brachyura*	63(12)
868.	BLUEWINGED PITTA *Pitta moluccensis*	63(11)
869.	HOODED or GREENBREASTED PITTA *Pitta sordida*	63(9)
871.	BLUE PITTA *Pitta cyanea*	63(8)

Family ALAUDIDAE: Larks

Small, dull-coloured sparrow-like birds, generally grey-brown and buff, cryptically patterned with brown or black. Terrestrial. Bill long and curved or short and stout. Legs short to long; hind claw straight, long and sharp. Wings long and pointed. Sexes generally alike. Flight strong; many species migratory; movement on ground by walking or running (*contra* hopping). Food, seeds and insects. Larks are well-known for their songs and soaring song-flights. Nest, a scrape on the ground with little or no lining. Eggs, 2-6.

		Plate
872.	SINGING BUSH LARK *Mirafra javanica*	64(1)
873.	BUSH LARK *Mirafra assamica*	64(2)
874.	Ssp. *affinis* of 873.	64(3)
877.	REDWINGED BUSH LARK *Mirafra erythroptera*	64(4)
878.	ASHYCROWNED FINCH-LARK *Eremopterix grisea*	64(5)
879.	BLACK CROWNED FINCH-LARK *Eremopterix nigriceps*	64(6)
880.	DESERT FINCH-LARK *Ammomanes deserti*	64(7)
881.	BARTAILED DESERT LARK *Ammomanes cincturus*	64(10)
882.	RUFOUSTAILED FINCH-LARK *Ammomanes phoenicurus*	64(9)
884.	BIFASCIATED or LARGE DESERT LARK *Alaemon alaudipes*	64(8)
886.	SHORT-TOED LARK *Calandrella cinerea*	64(14)
888a.	LESSER SHORT-TOED LARK *Calandrella rufescens*	64(13)
891.	SAND LARK *Calandrella raytal*	64(15)
892.	EASTERN CALANDRA LARK *Melanocorypha bimaculata*	64(11)
894.	LONGBILLED CALANDRA LARK *Melanocorypha maxima*	64(16)
895.	Ssp. *albigula* of 897	64(18)
897.	HORNED LARK *Eremophila alpestris*	64(17)
899.	CRESTED LARK *Galerida cristata*	64(20)
901.	MALABAR CRESTED LARK *Galerida malabarica*	64(21)
902.	SYKES'S CRESTED LARK *Galerida deva*	64(19)
903.	SKYLARK *Alauda arvensis*	64(23)
907.	EASTERN SYLARK *Alauda gulgula*	64(22)
EL.	CALANDRA LARK *Melanocorypha calandra*	64(12)

Family HIRUNDINIDAE: Swallows

Gregarious aerial birds, superficially similar to the swifts. Plumage chiefly black, brown, dark blue or dark green. Body slender; neck short. Bill short and flattened, with a very wide gape. Legs short; tarsi feathered in some species; feet small and weak but with strong claws. Wings long and pointed; tail medium to long, truncate to forked. Sexes nearly alike. Flight swift and strong. Food, mainly insects taken on the wing. Nest variable, of conglomerated mud pellets, usually attached to buildings or rocks; also burrows in vertical earth banks. Generally colonial. Eggs, 3-7. Young nidicolous; cared for by both sexes.

		Plate
910.	COLLARED SAND MARTIN *Riparia riparia*	65(2)
912.	PLAIN SAND MARTIN *Riparia paludicola*	65(3)
913.	CRAG MARTIN *Hirundo rupestris*	65(5)
914.	DUSKY CRAG MARTIN *Hirundo concolor*	65(6)

		Plate
915.	PALE CRAG MARTIN *Hirundo obsoleta*	65(4)
916.	SWALLOW *Hirundo rustica*	65(11)
918.	Ssp. *tytleri* of 916	65(12)
919.	HOUSE SWALLOW *Hirundo tahitica*	65(13)
921.	WIRETAILED SWALLOW *Hirundo smithii*	65(9)
922.	INDIAN CLIFF SWALLOW *Hirundo fluvicola*	65(17)
923.	STRIATED or REDRUMPED SWALLOW *Hirundo daurica*	65(14)
298.	Ssp. *hyperythra* of 923	65(15)
930.	HOUSE MARTIN *Delichon urbica*	65(10)
932.	NEPAL HOUSE MARTIN *Delichon nipalensis*	65(7)
EL.	ASIAN HOUSE MARTIN *Delichon dasypus*	65(8)

Family LANIIDAE: Shrikes

Bold and aggressive birds like miniature raptors with plumage chiefly grey or brown above and white below; face and flight feathers boldly marked in black and white. Head large. Bill strong and hooked. Legs and feet strong; claws sharp. Tail long, narrow and graduated. Strictly carnivorous, feeding on insects and small vertebrates. Known to maintain a 'larder' where surplus food is impaled on thorns to be eaten at leisure. This habit has given the shrikes the substantive name of 'butcher birds'. Voice, harsh calls as well as musical songs; most shrikes are accomplished mimics of other birds' calls. Nest, a deep cup of twigs, grass, wool or tow, etc., placed at moderate height in a thorn bush. Eggs, 3-6.

		Plate
933.	GREY SHRIKE *Lanius excubitor*	94(12)
937.	LESSER GREY SHRIKE *Lanius minor*	94(11)
938.	BURMESE SHRIKE *Lanius collurioides*	94(10)
940.	BAYBACKED SHRIKE *Lanius vittatus*	94(6)
941.	REDBACKED SHRIKE *Lanius collurio*	94(13)
942.	Ssp. *phoenicuroides* of 941	94(8)
943.	Ssp. *isabellinus* of 941	94(7)
945.	GREYBACKED or TIBETAN SHRIKE *Lanius tephronotus*	94(15)
946.	RUFOUSBACKED SHRIKE *Lanius schach*	94(17)
948.	Ssp. *tricolor* of 946	94(18)
949.	BROWN SHRIKE *Lanius cristatus*	94(9)
951.	WOODCHAT SHRIKE ?*Lanius senator*	94(14)

Family ORIOLIDAE: Orioles

Non-social arboreal birds. Plumage chiefly yellow, green, red, brown or black. Bill red, blue or black; strong, pointed and slightly hooked. Legs short and strong. Wings long and pointed. Sexes dimorphic. Flight rapid and undulating. Food, generally insects and fruit. Voice, loud fluty and melodious notes. Nest, a deep cup of woven grass and bast-fibres, slung like a hammock within a fork of branches. Eggs 2-5.

		Plate
952.	GOLDEN ORIOLE *Oriolus oriolus*	63(2)
954.	BLACKNAPED ORIOLE *Oriolus chinensis diffusus*	63(5)

		Plate
955.	Ssp. *tenuirostris* of 954	63(6)
958.	BLACKHEADED ORIOLE *Oriolus xanthornus*	63(4)
961.	MAROON ORIOLE *Oriolus traillii*	63(7)

Family DICRURIDAE: Drongos

Pugnacious arboreal birds. Plumage chiefly black or grey. Eyes reddish brown or red in most species. Bill stout, hooked and notched. Legs short, toes and claws stout, the latter curved and needle-sharp. Wings long; tail variable: medium to extremely long and deeply forked, some with projecting spatula-tipped 'wires'. Sexes alike. Flight swift and agile. Food, chiefly insects and other small creatures, and flower-nectar. Bold and fearless; will attack and drive away larger birds intruding the precincts of the nest-tree. Defenceless species often build nests in the same tree as drongos to take advantage of the birds' vigilance and pugnacity in warding off potential marauders. Nest, a shallow cup of twigs, grass, etc. compacted with cobwebs, placed in a horizontal fork of twigs. Eggs, 2-4.

		Plate
963.	BLACK DRONGO or KING-CROW *Dicrurus adsimilis*	69(1)
965.	GREY or ASHY DRONGO *Dicrurus leucophaeus*	69(8)
966a.	Ssp. *salangensis* of 965	69(9)
967.	WHITEBELLIED DRONGO *Dicrurus caerulescens*	69(12)
970.	CROWBILLED DRONGO *Dicrurus annectans*	69(3)
971.	BRONZED DRONGO *Dicrurus aeneus*	69(6)
972.	LESSER RACKET-TAILED DRONGO *Dicrurus remifer*	69(2)
973.	HAIRCRESTED or SPANGLED DRONGO *Dicrurus hottentottus*	69(7)
975.	ANDAMAN DRONGO *Dicrurus andamanensis*	69(10)
977.	GREATER RACKET-TAILED DRONGO *Dicrurus paradiseus*	69(4)
979.	Ssp. *lophorhinus* of 977	69(5)

Family ARTAMIDAE: Swallow-Shrikes or Wood Swallows

Arboreal birds with fine-textured soft plumage; the only passerines possessing powder-down feathers. Body stout; neck short. Bill stout, rather finch-like, slightly curved, pointed, and with a wide gape. Legs short and stout; feet strong. Wings long and pointed. Sexes nearly alike. Flight swift and swallow-like: several rapid wing-beats followed by an effortless, graceful glide. Food, insects caught on the wing. Nest, a loosely structured shallow cup of grass, roots, etc., placed on a branch, or in the head of a palm tree often at considerable heights from the ground. Extremely aggressive when nesting, attacking crows and raptors encroaching in the vicinity of the nest-tree. Eggs, 2-4.

		Plate
982.	ASHY SWALLOW-SHRIKE *Artamus fuscus*	65(18)
983.	WHITEBREASTED SWALLOW-SHRIKE *Artamus leucorhynchus*	65(16)

Family STURNIDAE: Starlings, Mynas

Gregarious arboreal or terrestrial birds with silky plumage. Bill typically straight and rather long and slender. Legs and feet strong. Wings short and rounded to long and pointed; tail usually short and square. Sexes nearly alike. Flight strong. Food, insects, fruits, grain and flower-nectar. Voice, a variety of harsh to pleasing notes and whistles. Some species are renowned mimics of the calls of other birds and human speech. Nest, in holes or cavities in trees. Eggs, 2-9. Some species have become commensals of man and some are serious pests of agriculture and horticulture.

		Plate
984.	SPOTTED WINGED STARE *Saroglossa spiloptera*	96(2)
986.	GLOSSY STARE or STARLING *Aplonis panayensis*	96(1)
987.	GREYHEADED MYNA *Sturnus malabaricus*	96(3)
988.	Ssp. *blythi* of 987	96(4)
991.	WHITEHEADED MYNA *Sturnus erythropygius*	96(8)
993.	CEYLON WHITEHEADED MYNA *Sturnus senex*	96(7)
994.	BLACKHEADED or BRAHMINY MYNA *Sturnus pagodarum*	96(5)
995.	DAURIAN MYNA *Sturnus sturninus*	96(6)
996.	ROSY PASTOR *Sturnus roseus*	96(10)
997.	STARLING *Sturnus vulgaris*	96(9)
1002.	PIED MYNA *Sturnus contra*	96(11)
1005.	CHINESE or GREYBACKED MYNA *Sturnus sinensis*	96(13)
1006.	COMMON MYNA *Acridotheres tristis*	96(12)
1008.	BANK MYNA *Acridotheres ginginianus*	96(14)
1009.	JUNGLE MYNA *Acridotheres fuscus*	96(15)
1012.	ORANGEBILLED JUNGLE MYNA *Acridotheres javanicus*	96(18)
1013.	COLLARED MYNA *Acridotheres albocinctus*	96(17)
1014.	GOLDCRESTED MYNA *Mino coronatus*	96(16)
1015.	GRACKLE or HILL MYNA *Gracula religiosa*	96(19)
1019.	CEYLON HILL MYNA Gracula ptilogenys	96(20)

Family CORVIDAE: Crows, Magpies, Jays, etc.

The largest passerine birds. Most species highly gregarious and social. Bill stout and powerful. Tarsi large, either fully booted or scutellated in front and booted behind. Wings and tail strong. Sexes alike or nearly so. Food, all types of plant and animal matter. Most species bold, and aggressive. Some species show a highly developed mentality and have a complex social organisation. Nest, generally bulky, open, made of twigs. Eggs, 3-6.

		Plate
1020.	JAY *Garrulus glandarius*	70(2)
1022.	BLACKTHROATED JAY *Garrulus lanceolatus*	70(1)
1023.	GREEN MAGPIE *Cissa chinensis*	70(4)
1024.	CEYLON MAGPIE *Cissa ornata*	70(5)

		Plate
1025.	YELLOWBILLED BLUE MAGPIE *Cissa flavirostris*	70(6)
1027.	REDBILLED BLUE MAGPIE *Cissa erythrorhyncha*	70(7)
1029.	WHITERUMPED MAGPIE *Pica pica*	70(3)
1032.	INDIAN TREE PIE *Dendrocitta vagabunda*	70(10)
1035.	BLACKBROWED TREE PIE *Dendrocitta frontalis*	70(11)
1036.	WHITEBELLIED TREE PIE *Dendrocitta leucogastra*	70(9)
1038.	HIMALAYAN TREE PIE *Dendrocitta formosae*	70(8)
1040.	ANDAMAN TREE PIE *Dendrocitta bayleyi*	70(12)
1041.	HUME'S GROUND CHOUGH *Podoces humilis*	70(15)
1042.	NUTCRACKER *Nucifraga caryocatactes*	70(13)
1043.	Ssp. *hemispila* of 1042	70(14)
1045.	YELLOWBILLED or ALPINE CHOUGH *Pyrrhocorax graculus*	71(2)
1046.	REDBILLED CHOUGH *Pyrrhocorax pyrrhocorax*	71(4)
1049.	HOUSE CROW *Corvus splendens*	71(6)
1052.	ROOK *Corvus frugilegus*	71(5)
1053.	JACKDAW *Corvus monedula*	71(3)
1054.	JUNGLE CROW *Corvus macrorhynchos*	71(7)
1058.	CARRION CROW *Corvus corone*	71(8)
1058a.	Ssp. *sharpii* of 1058	71(9)
1059.	RAVEN *Corvus corax*	71(11)
1061.	BROWN-NECKED RAVEN *Corvus ruficollis*	71(10)

Family BOMBYCILLIDAE: Waxwing, Silky Flycatcher

Rather sluggish arboreal birds with soft silky plumage. Crested. Bill short and thick, hooked, notched, and with a wide gape. Legs short. Wings short to long and pointed; tail short and square. The waxwings have the shafts of the secondaries prolonged into waxy tips. Food, berries, flowers, and insects caught on the wing. Flight rapid and direct or undulating. Nest, a cup of pliant twigs, moss and grass, in small trees and bushes. Eggs, 3-7.

		Plate
1062.	WAXWING *Bombycilla garrulus*	94(16)
1063.	GREY HYPOCOLIUS *Hypocolius ampelinus*	94(19)

Family CAMPEPHAGIDAE: Cuckoo-Shrikes and Minivets

Small to medium sized arboreal birds. Plumage soft, of loosely attached feathers, grey, black, blue, red, orange or yellow. Bill relatively heavy and slightly to strongly hooked. Legs short. Tail typically graduated and long. The birds rarely descend to the ground. Food, chiefly insects. Nest, a shallow open cup of twigs and rootlets cemented with cobweb on a high branch or in a fork of branches. Eggs, 2-5. Incubation by both sexes or by female alone.

		Plate
1065.	PIED FLYCATCHER-SHRIKE *Hemipus picatus*	66(12)
1068.	LARGE WOOD SHRIKE *Tephrodornis virgatus*	66(13)

		Plate
1070.	COMMON WOOD SHRIKE *Tephrodornis pondicerianus*	66(10)
1072.	LARGE CUCKOO-SHRIKE *Coracina novaehollandiae*	66(11)
1076.	BARRED CUCKOO-SHRIKE *Coracina striata*	66(17)
1077.	SMALLER GREY CUCKOO-SHRIKE *Coracina melaschistos*	66(16)
1078.	BLACKHEADED CUCKOO-SHRIKE *Coracina melanoptera*	66(14)
1079a.	PIED CUCKOO-SHRIKE *Coracina nigra*	66(15)
1080.	Ssp. *speciosus* of 1081	66(9)
1081.	SCARLET MINIVET *Pericrocotus flammeus*	66(8)
1084.	SHORTBILLED MINIVET *Pericrocotus brevirostris*	66(5)
1085.	LONGTAILED MINIVET *Pericrocotus ethologus*	66(6)
1088.	YELLOWTHROATED MINIVET *Pericrocotus solaris*	66(4)
1089.	ROSY MINIVET *Pericrocotus roseus*	66(1)
1089a.	ASHY MINIVET *Pericrocotus divaricatus*	66(7)
1093.	SMALL MINIVET *Pericrocotus cinnamomeus*	66(3)
1096.	WHITEBELLIED MINIVET *Pericrocotus erythropygius*	66(2)

Family IRENIDAE: Fairy Bluebird, Ioras and Leaf Birds

Small to medium arboreal birds of brightly coloured plumage, chiefly green, yellow, blue and black, or combinations. Bill moderately long and slightly hooked; in some species slender and curved. Legs short and thick; feet small. Wings rounded; tail short to long, square or rounded. Sexes dimorphic, the males more brightly coloured. Food, fruit, berries, seeds, flower nectar and insects. Voice, a variety of mellow and sibilant whistles, harsh chattering and mimicry (leaf birds). Nest, neat and cup-shaped, bound with cobweb built in trees. Eggs, 2-4.

		Plate
1098.	COMMON IORA *Aegithina tiphia*	67(1)
1102.	MARSHALL'S IORA *Aegithina nigrolutea*	67(3)
1103.	GOLDENFRONTED CHLOROPSIS or LEAF BIRD *Chloropsis aurifrons*	67(5)
1106.	ORANGEBELLIED CHLOROPSIS or LEAF BIRD *Chloropsis hardwickii*	67(4)
1108.	GOLDMANTLED CHLOROPSIS or LEAF BIRD *Chloropsis cochinchinensis*	67(6)
1109.	FAIRY BLUEBIRD *Irena puella*	67(14)

Family PYCNONOTIDAE: Bulbuls

Noisy, gregarious and mostly dull-coloured birds with soft, long and fluffy plumage. Bill short to medium length and slightly curved. Legs short and rather weak. Wings short and rounded; tail comparatively longer. Hairlike feathers on nape and rictal bristles usually well developed. Sexes generally alike. Food, fruit, nectar, berries, insects. Most have sprightly musical calls; some species renowned as good songsters. Nest, open cup of twigs, leaves, etc., generally built in bushes or trees. Eggs, 2-5.

		Plate
1111.	FINCHBILLED BULBUL *Spizixos canifrons*	68(1)
1112.	BLACKHEADED BULBUL *Pycnonotus atriceps*	68(7)
1114.	GREYHEADED BULBUL *Pycnonotus priocephalus*	68(3)
1115.	BLACKHEADED YELLOW BULBUL *Pycnonotus melanicterus flaviventris*	68(4)
1116.	RUBYTHROATED YELLOW BULBUL *Pycnonotus melanicterus gularis*	68(5)
1117.	BLACKCAPPED YELLOW BULBUL *Pycnonotus melanicterus melanicterus*	68(6)
1120.	REDWHISKERED BULBUL *Pycnonotus jocosus*	67(7)
1123.	Ssp. *leucotis* of 1125	67(11)
1125.	WHITECHEEKED BULBUL *Pycnonotus leucogenys*	67(12)
1128.	REDVENTED BULBUL *Pycnonotus cafer*	67(9)
1131.	Ssp. *bengalensis* of 1128	67(10)
1133.	STRIATED GREEN BULBUL *Pycnonotus striatus*	68(2)
1135.	YELLOWTHROATED BULBUL *Pycnonotus xantholaemus*	68(8)
1136.	YELLOWEARED BULBUL *Pycnonotus penicillatus*	68(10)
1137.	BLYTH'S BULBUL *Pycnonotus flavescens*	68(9)
1138.	WHITEBROWED BULBUL *Pycnonotus luteolus*	68(13)
1140.	WHITETHROATED BULBUL *Cringer flaveolus*	68(16)
1141.	OLIVE BULBUL *Hypsipetes viridescens*	68(12)
1142.	NICOBAR BULBUL *Hypsipetes nicobariensis*	68(14)
1144.	YELLOWBROWED BULBUL *Hypsipetes indicus*	68(15)
1146.	RUFOUSBELLIED BULBUL *Hypsipetes mcclellandi*	68(11)
1147.	BROWNEARED BULBUL *Hypsipetes flavalus*	67(8)
1148.	BLACK BULBUL *Hypsipetes madagascariensis*	67(13)

Family MUSCICAPIDAE: Babblers, Flycatchers, Warblers, Thrushes & Chats

A vast agglomeration of superficially heterogeneous passerine birds erstwhile recognised as four discrete families, namely Timaliidae (Babblers), Muscicapidae (Flycatchers), Sylviidae (Warblers), and Turdidae (Thrushes and Chats). The occurrence of connecting forms makes it impracticable to separate them from one another on clear-cut anatomical or evolutionary criteria. Recent behavioural studies, moreover, provide further confirmation of their close inter-relationship; hence, in conformity with international consensus the families have been demoted to subfamilial rank and merged together into a single huge family. However, for greater convenience the four subfamilies and the contained Indian species of each are here dealt with individually.

Subfamily TIMALIINAE; Babblers

Arboreal or terrestrial birds with soft lax plumage. Bill generally strong. Legs and feet strong. Wings short and rounded. Flight weak. Food, insects, other small animals, flower-nectar, and fruit. Most species noisy, with harsh calls; some with melodious song. Nest, cup-shaped or domed, built chiefly on trees and bushes. Eggs, 1-7.

		Plate
1154.	SPOTTED BABBLER *Pellorneum ruficeps*	75(20)
1160.	MARSH SPOTTED BABBLER *Pellorneum palustre*	75(19)
1161.	BROWNCAPPED BABBLER *Pellorneum fuscocapillum*	75(16)
1164.	BROWN BABBLER *Pellorneum albiventre*	75(15)
1166.	TICKELL'S BABBLER *Trichastoma tickelli*	75(17)
1167.	ABBOTT'S BABBLER *Trichastoma abbotti*	75(18)
1169.	Ssp. *schisticeps* of 1173	74(18)
1173.	SLATYHEADED SCIMITAR BABBLER *Pomatorhinus horsfieldii*	74(19)
1178.	RUFOUSNECKED SCIMITAR BABBLER *Pomatorhinus ruficollis*	74(16)
1181.	RUSTYCHEEKED SCIMITAR BABBLER *Pomatorhinus erythrogenys*	74(20)
1185.	LARGE SCIMITAR BABBLER *Pomatorhinus hypoleucos*	74(15)
1186.	CORALBILLED SCIMITAR BABBLER *Pomatorhinus ferruginosus*	74(13)
1187.	Ssp. *formosus* of 1186	74(12)
1189.	LONGBILLED SCIMITAR BABBLER *Pomatorhinus ochraceiceps*	74(17)
1191.	SLENDERBILLED SCIMITAR BABBLER *Xiphirhynchus superciliaris*	74(14)
1193.	LONGBILLED WREN BABBLER *Rimator malacoptilus*	75(9)
1194.	STREAKED or SHORT-TAILED WREN-BABBLER *Napothera brevicaudata*	75(12)
1195.	SMALL WREN-BABBLER *Napothera epilepidota*	75(11)
1198.	SCALYBREASTED WREN-BABBLER *Pnoepyga albiventer*	75(5)
1199.	BROWN or LESSER SCALYBREASTED WREN-BABBLER *Pnoepyga pusilla*	75(3)
1200.	TAILED WREN-BABBLER *Spelaeornis caudatus*	75(4)
1201.	MISHMI WREN *Spelaeornis badeigularis*	75(8)
1202.	LONGTAILED WREN-BABBLER *Spelaeornis longicaudatus*	75(13)
1203.	STREAKED LONGTAILED WREN-BABBLER *Spelaeornis chocolatinus*	75(7)
1205.	LONGTAILED SPOTTED WREN-BABBLER *Spelaeornis troglodytoides*	75(10)
1206.	SPOTTED WREN-BABBLER *Spelaeornis formosus*	75(6)
1207.	WEDGEBILLED WREN *Sphenocichla humei*	75(14)
1209.	REDFRONTED BABBLER *Stachyris rufifrons*	74(1)
1210.	REDHEADED BABBLER *Stachyris ruficeps*	74(5)
1211.	REDBILLED BABBLER *Stachyris pyrrhops*	74(2)
1212.	GOLDHEADED BABBLER *Stachyris chrysaea*	74(3)
1214.	BLACKTHROATED BABBLER *Stachyris nigriceps*	74(9)
1218.	AUSTEN'S SPOTTED BABBLER *Stachyris oglei*	74(8)
1219.	Ssp. *abuensis* of 1222	74(7)

		Plate
1222.	RUFOUSBELLIED BABBLER *Dumetia hyperythra*	74(6)
1224.	BLACKHEADED BABBLER *Rhopocichla atriceps*	74(10)
1226.	Ssp. *siccatus* of 1224	74(11)
1228.	YELLOWBREASTED BABBLER *Macronous gularis*	74(4)
1229.	REDCAPPED BABBLER *Timalia pileata*	79(1)
1231.	YELLOWEYED BABBLER *Chrysomma sinense*	79(2)
1233.	JERDON'S BABBLER *Chrysomma altirostre*	79(3)
1235.	BEARDED TIT-BABBLER or REEDLING *Panurus biarmicus*	73(15)
1236.	GREAT PARROTBILL *Conostoma aemodium*	79(15)
1237.	BROWN SUTHORA or PARROTBILL *Paradoxornis unicolor*	79(13)
1238.	FULVOUSFRONTED SUTHORA or PARROTBILL *Paradoxornis fulvifrons*	73(24)
1241.	ORANGE SUTHORA or PARROTBILL *Paradoxornis nipalensis humii*	73(22)
1242.	ORANGE SUTHORA or PARROTBILL *Paradoxornis nipalensis poliotis*	73(23)
1245.	Ssp. *oatesi* of 1246	73(20)
1246.	LESSER REDHEADED SUTHORA or PARROTBILL *Paradoxornis atrosuperciliaris*	73(21)
1247.	GREATER REDHEADED PARROTBILL *Paradoxornis ruficeps*	73(19)
1249.	GREYHEADED PARROTBILL *Paradoxornis gularis*	73(16)
1251.	BLACKTHROATED PARROTBILL *Paradoxornis flavirostris*	73(17)
1252.	WHITETHROATED PARROTBILL *Paradoxornis guttaticollis*	73(18)
1254.	COMMON BABBLER *Turdoides caudatus*	79(4)
1256.	STRIATED BABBLER *Turdoides earlei*	79(5)
1257.	SLENDERBILLED BABBLER *Turdoides longirostris*	79(10)
1258.	LARGE GREY BABBLER *Turdoides malcolmi*	79(8)
1259.	RUFOUS BABBLER *Turdoides subrufus*	79(12)
1265.	JUNGLE BABBLER *Turdoides striatus*	79(7)
1266.	CEYLON RUFOUS BABBLER *Turdoides rufescens*	79(11)
1267.	WHITEHEADED BABBLER *Turdoides affinis*	79(9)
1269.	SPINY BABBLER *Turdoides nipalensis*	79(6)
1270.	CHINESE BABAX *Babax lanceolatus*	76(12)
1271.	GIANT TIBETAN BABAX *Babax waddelli*	76(14)
1272.	ASHYHEADED LAUGHING THRUSH *Garrulax cinereifrons*	79(14)
1274.	WHITETHROATED LAUGHING THRUSH *Garrulax albogularis*	77(15)
1275.	NECKLACED LAUGHING THRUSH *Garrulax moniligerus*	76(11)
1277.	BLACKGORGETED LAUGHING THRUSH *Garrulax pectoralis*	76(13)
1279.	STRIATED LAUGHING THRUSH *Garrulax striatus*	76(10)
1283.	WHITECRESTED LAUGHING THRUSH *Garrulax leucolophus*	77(16)
1285.	CHESTNUTBACKED LAUGHING THRUSH *Garrulax nuchalis*	77(10)
1286.	YELLOWTHROATED LAUGHING THRUSH *Garrulax galbanus*	77(5)
1287.	RUFOUSVENTED LAUGHING THRUSH *Garrulax delesserti*	77(6)
1288.	Ssp. *gularis* of 1287.	77(7)
1290.	VARIEGATED LAUGHING THRUSH *Garrulax ariegatus*	77(13)
1291.	ASHY LAUGHING THRUSH *Garrulax cineraceus*	77(4)
1294.	RUFOUSCHINNED LAUGHING THRUSH *Garrulax rufogularis*	76(9)

		Plate
1297.	Ssp. *maximus* of 1299	76(15)
1299.	WHITESPOTTED LAUGHING THRUSH *Garrulax ocellatus*	76(16)
1300.	GREYSIDED LAUGHING THRUSH *Garrulax caerulatus*	77(14)
1303.	RUFOUSNECKED LAUGHING THRUSH *Garrulax ruficollis*	77(9)
1304.	SPOTTEDBREASTED LAUGHING THRUSH *Garrulax merulinus*	76(8)
1306.	WHITEBROWED LAUGHING THRUSH *Garrulax sannio*	77(8)
1307.	NILGIRI LAUGHING THRUSH *Garrulax cachinnans*	77(2)
1309.	WHITEBREASTED LAUGHING THRUSH *Garrulax jerdoni*	77(3)
1314.	STREAKED LAUGHING THRUSH *Garrulax lineatus*	76(2)
1317.	MANIPUR STREAKED LAUGHING THRUSH *Garrulax virgatus*	76(7)
1318.	BROWNCAPPED LAUGHING THRUSH *Garrulax austeni*	76(6)
1319.	BLUEWINGED LAUGHING THRUSH *Garrulax squamatus*	76(5)
1320.	PLAINCOLOURED LAUGHING THRUSH *Garrulax subunicolor*	76(1)
1321.	PRINCE HENRI'S LAUGHING THRUSH *Garrulax henrici*	77(11)
1322.	BLACKFACED LAUGHING THRUSH *Garrulax affinis*	77(12)
1324.	REDHEADED LAUGHING THRUSH *Garrulax erythrocephalus*	76(4)
1326.	Ssp. *nigrimentum* of 1324	76(3)
1331.	CRIMSONWINGED LAUGHING THRUSH *Garrulax phoeniceus*	77(1)
1333.	SILVEREARED MESIA *Leiothrix argentauris*	80(1)
1336.	REDBILLED LEIOTHRIX *Leiothrix lutea*	80(2)
1338.	FIRETAILED MYZORNIS *Myzornis pyrrhoura*	67(2)
1339.	NEPAL CUTIA *Cutia nipalensis*	80(3)
1340.	RUFOUSBELLIED SHRIKE-BABBLER *Pteruthius rufiventer*	80(4)
1341.	REDWINGED SHRIKE-BABBLER *Pteruthius flaviscapis*	80(8)
1343.	GREEN SHRIKE-BABBLER *Pteruthius xanthochlorus*	80(5)
1345.	CHESTNUT-THROATED SHRIKE-BABBLER *Pteruthius melanotis*	80(7)
1346.	CHESTNUTFRONTED SHRIKE-BABBLER *Pteruthius aenobarbus*	80(6)
1347.	WHITEHEADED SHRIKE-BABBLER *Gampsorhynchus rufulus*	78(1)
1348.	SPECTACLED BARWING *Actinodura egertoni*	78(4)
1352.	HOARY BARWING *Actinodura nipalensis*	78(2)
1355.	AUSTEN'S BARWING *Actinodura nipalensis waldeni*	78(5)
1357.	REDTAILED MINLA *Minla ignotincta*	80(12)
1359.	BARTHROATED SIVA *Minla strigula*	80(9)
1362.	BLUEWINGED SIVA *Minla cyanouroptera*	80(10)
1365.	WHITEBROWED YUHINA *Yuhina castaniceps*	78(13)
1366.	WHITENAPED YUHINA *Yuhina bakeri*	78(19)
1368.	YELLOWNAPED YUHINA *Yuhina flavicollis*	78(15)
1372.	STRIPETHROATED YUHINA *Yuhina gularis*	78(18)
1373.	RUFOUSVENTED YUHINA *Yuhina occipitalis*	78(17)
1374.	BLACKCHINNED YUHINA *Yuhina nigrimenta*	78(16)
1375.	WHITEBELLIED YUHINA *Yuhina xantholeuca*	78(12)
1376.	GOLDENBREASTED TIT-BABBLER *Alcippe chrysotis*	80(11)
1378.	DUSKY-GREEN or YELLOWTHROATED TIT-BABBLER *Alcippe cinerea*	78(7)
1379.	CHESTNUT-HEADED TIT-BABBLER *Alcippe castaneceps*	78(6)
1381.	WHITEBROWED TIT-BABBLER *Alcippe vinipectus*	78(14)
1385.	GREY-HEADED TIT-BABBLER *Alcippe cinereiceps*	78(10)

		Plate
1386.	REDTHROATED TIT-BABBLER *Alcippe rufogularis*	78(8)
1388.	RUFOUSHEADED TIT-BABBLER *Alcippe brunnea*	78(3)
1390.	QUAKER BABBLER *Alcippe poioicephala*	78(11)
1392.	NEPAL BABBLER *Alcippe nipalensis*	78(9)
1395.	CHESTNUTBACKED SIBIA *Heterophasia annectens*	80(14)
1396.	BLACKCAPPED SIBIA *Heterophasia capistrata*	80(13)
1399.	GREY SIBIA *Heterophasia gracilis*	80(17)
1400.	BEAUTIFUL SIBIA *Heterophasia pulchella*	80(15)
1401.	LONGTAILED SIBIA *Heterophasia picaoides*	80(16)
EL.	SPOTBREASTED SCIMITAR BABBLER *Pomatorhinus erythrocnemis*	74(21)

Subfamily MUSCICAPINAE: Flycatchers

Dull to brightly coloured small to medium-sized arboreal birds. Bill typically broad and flat at base. Legs short. Wings short and rounded to long and pointed. Tail variable: short and square or notched; or graduated and fan-shaped, to longish with flowing ribbons. Food, insects and spiders, often caught on the wing. Some species have well-developed songs. Nest, a neatly built deepish cup of grasses etc. usually bound or plastered with cobweb, placed on a tree or bush, sometimes in holes. Eggs, 2-6. Incubation and nest-feeding by both sexes, but female does the most work.

		Plate
1402.	OLIVE FLYCATCHER *Rhinomyias brunneata*	92(13)
1403.	SPOTTED FLYCATCHER *Muscicapa striata*	92(5)
1406.	SOOTY FLYCATCHER *Muscicapa sibirica*	92(7)
1407.	BROWN FLYCATCHER *Muscicapa latirostris*	92(11)
1408.	BROWNBREASTED FLYCATCHER *Muscicapa muttui*	92(6)
1409.	RUFOUSTAILED FLYCATCHER *Muscicapa ruficauda*	92(8)
1410.	FERRUGINOUS FLYCATCHER *Muscicapa ferruginea*	92(9)
1411.	REDBREASTED FLYCATCHER *Muscicapa parva*	92(2)
1412.	Ssp. *albicilla* of 1411	92(3)
1413.	KASHMIR REDBREASTED FLYCATCHER *Muscicapa subrubra*	92(4)
1414.	ORANGEGORGETED FLYCATCHER *Muscicapa strophiata*	92(12)
1415.	WHITEGORGETED FLYCATCHER *Muscicapa monileger*	92(21)
1416.	Ssp. *leucops* of 1415	92(20)
1417.	RUFOUSBREASTED BLUE FLYCATCHER *Muscicapa hyperythra*	93(5)
1418.	RUSTYBREASTED BLUE FLYCATCHER *Muscicapa hodgsonii*	93(4)
1419,	1420. LITTLE PIED FLYCATCHER *Muscicapa westermanni*	92(14)
1421.	WHITEBROWED BLUE FLYCATCHER *Muscicapa superciliaris*	92(15)
1423.	SLATY BLUE FLYCATCHER *Muscicapa leucomelanura*	92(18)
1424.	Ssp. *minuta* of 1423	92(17)
1426.	SAPPHIREHEADED FLYCATCHER *Muscicapa sapphira*	93(2)
1427.	BLACK-AND-ORANGE FLYCATCHER *Muscicapa nigrorufa*	92(10)
1428.	LARGE NILTAVA *Muscicapa grandis*	92(19)
1429.	SMALL NILTAVA *Muscicapa macgrigoriae*	92(16)
1432.	RUFOUSBELLIED NILTAVA *Muscicapa sundara*	93(14)

		Plate
1433.	RUFOUSBELLIED BLUE FLYCATCHER *Muscicapa vivida*	93(16)
1434.	WHITETAILED BLUE FLYCATCHER *Muscicapa concreta*	93(15)
1435.	WHITEBELLIED BLUE FLYCATCHER *Muscicapa pallipes*	93(9)
1436.	BROOKS'S FLYCATCHER *Muscicapa poliogenys*	93(8)
1439.	PALE BLUE FLYCATCHER *Muscicapa unicolor*	93(11)
1440.	BLUETHROATED FLYCATCHER *Muscicapa rubeculoides*	93(6)
1441.	LARGEBILLED BLUE FLYCATCHER *Muscicapa banyumas*	93(7)
1442.	TICKELL'S BLUE FLYCATCHER *Muscicapa tickelliae*	93(1)
1444.	DUSKY BLUE FLYCATCHER *Muscicapa sordida*	93(12)
1445.	VERDITER FLYCATCHER *Muscicapa thalassina*	93(10)
1446.	NILGIRI FLYCATCHER *Muscicapa albicaudata*	93(13)
1447.	PIGMY BLUE FLYCATCHER *Muscicapella hodgsoni*	93(3)
1449.	GREYHEADED FLYCATCHER *Culicicapa ceylonensis*	91(8)
1450.	YELLOWBELLIED FANTAIL FLYCATCHER *Rhipidura hypoxantha*	91(1)
1451.	WHITEBROWED FANTAIL FLYCATCHER *Rhipidura aureola*	94(4)
1455.	WHITETHROATED FANTAIL FLYCATCHER *Rhipidura albicollis*	94(5)
1458.	Ssp. *albogularis* of 1455	94(3)
1461.	PARADISE FLYCATCHER *Terpsiphone paradisi*	94(2)
1465.	BLACKNAPED FLYCATCHER *Hypothymis azurea*	94(1)

Subfamily PACHYCEPHALINAE: Thickheads or Shrikebilled Flycatchers

Bill strong, deep, laterally compressed, notched at tip. Three strong rictal bristles; numerous smaller hairs overhanging nostril. Wing long; tail square. Sexes alike.

		Plate
1470.	GREY THICKHEAD or MANGROVE WHISTLER *Pachycephala grisola*	92(1)

Subfamily SYLVIINAE: Warblers

Small, dull-coloured, active arboreal birds. Bill slender and pointed. Legs short. Wings of medium length and rounded. Food, insects, spiders, etc., sometimes berries. Most species have well-developed melodious songs that are often diagnostic for species otherwise confusingly alike. Nest, cup-shaped or domed, placed on the ground among vegetation or up in trees or bushes. Eggs, 3-5 or 6.

		Plate
1471.	DULL SLATYBELLIED GROUND WARBLER*Tesia cyaniventer*	91(15)
1472.	SLATYBELLIED GROUND WARBLER *Tesia olivea*	91(16)
1473.	CHESTNUT-HEADED GROUND WARBLER *Tesia castaneocoronata*	91(17)
1474.	PALEFOOTED BUSH WARBLER *Cettia pallidipes*	87(1)
1476.	JAPANESE BUSH WARBLER *Cettia diphone*	87(2)
1478.	STRONGFOOTED BUSH WARBLER *Cettia montana*	87(3)
1479.	LARGE BUSH WARBLER *Cettia major*	87(6)
1481,	1580. ABERRANT BUSH WARBLER *Cettia flavolivacea*	87(4), 90(7)

		Plate
1484.	HUME'S BUSH WARBLER *Cettia acanthizoides*	87(7)
1486.	RUFOUSCAPPED BUSH WARBLER *Cettia brunnifrons*	87(5)
1488.	CETTI'S WARBLER *Cettia cetti*	87(11)
1490.	SPOTTED BUSH WARBLER *Bradypterus thoracicus*	87(10)
1491.	LARGEBILLED BUSH WARBLER *Bradypterus major*	87(13)
1492.	CHINESE BUSH WARBLER *Bradypterus tacsanowskius*	87(15)
1493.	BROWN BUSH WARBLER *Bradypterus luteoventris*	87(8)
1494.	PALLISER'S WARBLER *Bradypterus palliseri*	87(9)
1495.	MOUSTACHED SEDGE WARBLER *Acrocephalus melanopogon*	89(10)
1496.	FANTAIL WARBLER *Cisticola exilis*	89(22)
1498.	STREAKED FANTAIL WARBLER *Cisticola juncidis*	89(20)
1501.	BEAVAN'S WREN-WARBLER *Prinia rufescens*	88(16)
1503.	ASHY-GREY WREN-WARBLER *Prinia hodgsonii*	88(17)
1506.	RUFOUSFRONTED WREN-WARBLER *Prinia buchanani*	89(8)
1507.	HODGSON'S WREN-WARBLER *Prinia cinereocapilla*	88(15)
1508.	STREAKED WREN-WARBLER *Prinia gracilis*	89(7)
1511.	PLAIN WREN-WARBLER *Prinia subflava*	89(12)
1515.	Ssp. *stewarti* of 1517	88(14)
1517.	ASHY WREN-WARBLER *Prinia socialis*	88(13)
1521.	JUNGLE WREN-WARBLER *Prinia sylvatica*	89(13)
1524.	Ssp. *sindiana* of 1525	89(14)
1525.	YELLOWBELLIED WREN-WARBLER *Prinia flaviventris*	89(15)
1527.	BROWN HILL WARBLER *Prinia criniger*	89(19)
1529.	BLACKTHROATED HILL WARBLER *Prinia atrogularis*	89(11)
1531.	LONGTAILED GRASS WARBLER *Prinia burnesii*	89(17)
1532.	Ssp. *cinerascens* of 1531	89(16)
1533.	STREAKED SCRUB WARBLER *Scotocerca inquieta*	89(18)
1534.	LARGE GRASS WARBLER *Graminicola bengalensis*	89(21)
1538.	TAILOR BIRD *Orthotomus sutorius*	91(11)
1540.	BLACKNECKED TAILOR BIRD *Orthotomus atrogularis*	91(14)
1541.	GOLDENHEADED TAILOR BIRD *Orthotomus cucullatus*	91(12)
1543.	PALLAS'S GRASSHOPPER WARBLER *Locustella certhiola*	89(1)
1544.	STREAKED GRASSHOPPER WARBLER *Locustella lanceolata*	89(9)
1545.	GRASSHOPPER WARBLER *Locustella naevia*	89(3)
EL.	Ssp. *obscurior* of 1545	89(4)
1546.	BROADTAILED GRASS WARBLER *Schoenicola platyura*	89(2)
1547.	BRISTLED GRASS WARBLER *Chaetornis striatus*	89(6)
1548.	STRIATED MARSH WARBLER *Megalurus palustris*	89(5)
1549.	THICKBILLED WARBLER *Acrocephalus aedon*	87(14)
1550.	INDIAN GREAT REED WARBLER *Acrocephalus stentoreus*	87(19)
1555.	BLACKBROWED REED WARBLER *Acrocephalus bistrigiceps*	87(12)
1555a.	REED WARBLER *Acrocephalus scirpaceus*	87(18)
1556.	BLYTH'S REED WARBLER *Acrocephalus dumetorum*	87(16)
1557.	PADDYFIELD WARBLER *Acrocephalus agricola*	87(17)
1562.	BOOTED WARBLER *Hippolais caligata*	90(3)
1564.	UPCHER'S WARBLER *Hippolais languida*	90(1)
1564a.	BARRED WARBLER *Sylvia nisoria*	88(1)

Subfamily TURDINAE: Thrushes and Chats

Plump, arboreal or terrestrial birds with soft plumage. Bill slender. Legs and feet stout; tarsi booted. Wings short and rounded to long and pointed; tail square, rounded or emarginate. Food, a variety of animal and vegetable matter. Some species have highly developed songs. Nest, an open cup in trees, bushes, on the ground, or in holes. Eggs, 2-6.

		Plate
1635.	GOULD'S SHORTWING *Brachypteryx stellata*	81(1)
1636.	RUSTYBELLIED SHORTWING *Brachypteryx hyperythra*	81(2)
1637.	RUFOUSBELLIED SHORTWING *Brachypteryx major*	81(4)
1639.	LESSER SHORTWING *Brachypteryx leucophrys*	81(3)
1640.	WHITEBROWED SHORTWING *Brachypteryx montana*	81(5)
1641.	RUFOUS CHAT *Erythropygia galactotes*	82(1)
1642.	NIGHTINGALE *Erithacus megarhynchos*	81(6)
1643.	RUBYTHROAT *Erithacus calliope*	81(7)
1644.	BLUETHROAT *Erithacus svecicus*	81(9)
1647.	HIMALAYAN RUBYTHROAT *Erithacus pectoralis*	81(8)
1650.	BLUE CHAT *Erithacus brunneus*	81(11)
1652.	FIRETHROAT *Erithacus pectardens*	81(10)
1653.	SIBERIAN BLUE CHAT *Erithacus cyane*	81(12)
1656.	ORANGEFLANKED BUSH ROBIN *Erithacus cyanurus*	81(14)
1658.	GOLDEN BUSH ROBIN *Erithacus chrysaeus*	81(15)
1659.	WHITEBROWED BUSH ROBIN *Erithacus indicus*	81(13)
1660.	RUFOUSBELLIED BUSH ROBIN *Erithacus hyperythrus*	81(17)
1661.	MAGPIE-ROBIN or DHYAL *Copsychus saularis*	83(4)
1665.	SHAMA *Copsychus malabaricus malabaricus*	83(1)
1668.	ANDAMAN SHAMA *Copsychus malabaricus albiventris*	83(2)
1669.	EVERSMANN'S REDSTART *Phoenicurus erythronotus*	82(2)
1670.	BLUEHEADED REDSTART *Phoenicurus caeruleocephalus*	82(5)
1671.	BLACK REDSTART *Phoenicurus ochruros phoenicuroides*	82(4)
1672.	BLACK REDSTART *Phoenicurus ochruros rufiventris*	82(3)
1673.	REDSTART *Phoenicurus phoenicurus*	82(6)
1674.	HODGSON'S REDSTART *Phoenicurus hodgsoni*	82(7)
1675.	BLUEFRONTED REDSTART *Phoenicurus frontalis*	82(8)
1676.	WHITETHROATED REDSTART *Phoenicurus schisticeps*	82(9)
1677.	DAURIAN REDSTART *Phoenicurus auroreus*	82(11)
1678.	GULDENSTÄDT'S REDSTART *Phoenicurus erythrogaster*	82(10)
1679.	PLUMBEOUS REDSTART *Rhyacornis fuliginosus*	82(12)
1680.	HODGSON'S SHORTWING *Hodgsonius phoenicuroides*	82(13)
1681.	WHITETAILED BLUE ROBIN *Cinclidium leucurum*	82(14)
1682.	BLUEFRONTED ROBIN *Cinclidium frontale*	82(15)
1683.	HODGSON'S GRANDALA *Grandala coelicolor*	83(3)
1684.	LITTLE FORKTAIL *Enicurus scouleri*	83(5)
1685.	BLACKBACKED FORKTAIL *Enicurus immaculatus*	83(7)
1686.	SLATYBACKED FORKTAIL *Enicurus schistaceus*	83(6)
1687.	LESCHENAULT'S FORKTAIL *Enicurus leschenaulti*	83(9)
1688.	SPOTTED FORKTAIL *Enicurus maculatus*	83(8)

		Plate
1690.	PURPLE COCHOA *Cochoa purpurea*	83(11)
1691.	GREEN COCHOA *Cochoa viridis*	83(10)
1692.	BROWN ROCK CHAT *Cercomela fusca*	82(18)
1693.	STOLICZKA'S BUSH CHAT *Saxicola macrorhyncha*	84(1)
1694.	HODGSON'S BUSH CHAT *Saxicola insignis*	84(4)
1696.	Ssp. *przewalskii* of 1697	84(3)
1697.	STONE CHAT *Saxicola torquata*	84(2)
1699.	WHITETAILED STONE CHAT *Saxicola leucura*	84(7)
1700.	PIED BUSH CHAT *Saxicola caprata*	84(5)
1704.	JERDON'S BUSH CHAT *Saxicola jerdoni*	84(6)
1705.	DARK-GREY BUSH CHAT *Saxicola ferrea*	84(8)
1706.	ISABELLINE CHAT *Oenanthe isabellina*	84(10)
1707.	REDTAILED CHAT *Oenanthe xanthoprymna*	84(11)
1708.	WHEATEAR *Oenanthe oenanthe*	84(9)
1710.	DESERT WHEATEAR *Oenanthe deserti*	84(12)
1711.	BARNES'S CHAT *Oenanthe finschi*	84(15)
1712.	PIED CHAT *Oenanthe picata*	84(17)
1713.	HOODED CHAT *Oenanthe monacha*	84(16)
1714.	HUME'S CHAT *Oenanthe alboniger*	84(14)
1715.	PLESCHANKA'S PIED CHAT or WHEATEAR *Oenanthe pleschanka*	84(13)
1716.	WHITECAPPED REDSTART or RIVER CHAT *Chaimarrornis leucocephalus*	82(19)
1717.	Ssp. *cambaiensis* of 1720	82(17)
1720.	INDIAN ROBIN *Saxicoloides fulicata*	82(16)
1722.	ROCK THRUSH *Monticola saxatilis*	85(7)
1723.	BLUEHEADED ROCK THRUSH *Monticola cinclorhynchus*	85(5)
1724.	CHESTNUTBELLIED ROCK THRUSH *Monticola rufiventris*	85(1)
1726.	BLUE ROCK THRUSH *Monticola solitarius*	85(4)
1727.	CEYLON WHISTLING THRUSH *Myiophonus blighi*	83(12)
1728.	MALABAR WHISTLING THRUSH *Myiophonus horsfieldii*	83(13)
1729.	BLUE WHISTLING THRUSH *Myiophonus caeruleus*	83(14)
1731.	PIED GROUND THRUSH *Zoothera wardii*	86(1)
1732.	SIBERIAN GROUND THRUSH *Zoothera sibirica*	86(3)
1733.	ORANGEHEADED GROUND THRUSH *Zoothera citrina*	85(2)
1734.	Ssp. *cyanotus* of 1733	85(3)
1737.	SPOTTED WINGED GROUND THRUSH *Zoothera spiloptera*	86(2)
1739.	PLAINBACKED MOUNTAIN THRUSH *Zoothera mollissima*	86(4)
1740.	LONGTAILED MOUNTAIN THRUSH *Zoothera dixoni*	86(6)
1741.	GOLDEN or SMALLBILLED MOUNTAIN THRUSH *Zoothera dauma*	86(8)
1745.	LARGE BROWN THRUSH *Zoothera monticola*	86(7)
1746.	LESSER BROWN THRUSH *Zoothera marginata*	86(5)
1747.	BLACKBREASTED THRUSH *Turdus dissimilis*	86(10)
1748.	TICKELL'S THRUSH *Turdus unicolor*	85(6)
1749.	WHITECOLLARED BLACKBIRD *Turdus albocinctus*	85(9)
1750.	GREYWINGED BLACKBIRD *Turdus boulboul*	85(8)

		Plate
1752.	BLACKBIRD *Turdus merula maximus*	85(13)
1753,	1754. BLACKBIRD *Turdus merula nigropileus*	85(11)
1755.	BLACKBIRD *Turdus merula simillimus*	85(12)
1757.	BLACKBIRD *Turdus merula kinnisii*	85(10)
1758.	GREYHEADED THRUSH *Turdus rubrocanus*	85(16)
1759.	Ssp. *gouldii* of 1758	85(17)
1760.	KESSLER'S THRUSH *Turdus kessleri*	85(15)
1761.	FEA'S THRUSH *Turdus feai*	85(14)
1762.	DARK THRUSH *Turdus obscurus*	86(11)
1763.	BLACKTHROATED THRUSH *Turdus ruficollis atrogularis*	86(13)
1764.	REDTHROATED THRUSH *Turdus ruficollis ruficollis*	86(12)
1765.	DUSKY THRUSH *Turdus naumanni*	86(9)
1766.	FIELDFARE *Turdus pilaris*	86(15)
1767.	REDWING *Turdus illiacus*	86(14)
1768.	MISTLE THRUSH *Turdus viscivorus*	86(16)
EL.	WHITETHROATED ROBIN *Irania gutturalis*	81(16)

Family TROGLODYTIDAE: Wrens

Tiny, skulking terrestrial birds with plumage typically 'pepper-and-salt' patterned in shades with white, buff, and black. Bill slender and medium to long and curved. Legs and feet strong; anterior toes partly adherent; claws long. Wings short and rounded; tail very short to longish. Sexes nearly alike. Food, chiefly insects. Known for their loud, spirited, melodious songs. Nest, in cavities in trees, rocks, buildings, etc., or domed nests built in bushes and trees. Polygamy common in many species.

		Plate
1770.	WREN *Troglodytes troglodytes*	75(1)
1771.	Ssp. *nipalensis* of 1770.	75(2)

Family CINCLIDAE: Dippers

Solitary aquatic birds inhabiting rapid-flowing mountain streams. Body compact and plumage firm and dense. Bill longish, slender, laterally compressed, slightly hooked and notched. Legs and toes long and stout; claws short and strong. Wings short, pointed and concave beneath; tail short, square or rounded. Sexes alike. Food, chiefly aquatic insects and other invertebrates. Walk and 'fly' under water, using the wings for propulsion. Nest, domed, built near water, lined with moss and leaves. Eggs, 4-5.

		Plate
1773.	WHITEBREASTED DIPPER *Cinclus cinclus*	69(13)
1775.	BROWN DIPPER *Cinclus pallasii*	69(14)

Family PRUNELLIDAE: Accentors or 'Hedge Sparrows'

Small, terrestrial sparrow-like birds. Plumage black, brown, grey or buff, streaked above and plain or streaked below. Bill slender and finely pointed. Legs and feet strong. Wings short and rounded to long and pointed. Sexes alike or nearly so. Food, chiefly insects in summer; fruits and berries in winter. Nest, neat and cup-shaped, built in trees or bushes or in a rock crevice. Eggs, 3-5.

		Plate
1777 to 1779	ALPINE ACCENTOR *Prunella collaris*	100(1)
1780.	ALTAI ACCENTOR *Prunella himalayana*	100(5)
1781.	ROBIN ACCENTOR *Prunella rubeculoides*	100(3)
1783.	RUFOUSBREASTED ACCENTOR *Prunella strophiata*	100(2)
1784.	BROWN ACCENTOR *Prunella fulvescens*	100(9)
1785a.	(RADDE'S) BROWN ACCENTOR *Prunella fulvescens ocularis*	100(7)
1787.	BLACKTHROATED ACCENTOR *Prunella atrogularis*	100(6)
1787a.	SIBERIAN ACCENTOR *Prunella montanella*	100(8)
1788.	MAROONBACKED ACCENTOR *Prunella immaculata*	100(4)

Family PARIDAE: Tits or Titmice

Small, restless, active, arboreal woodland birds. Plumage chiefly olive, brown, grey, white or buff. Bill stout, conical or pointed. Legs short and strong. Wings rounded. Food, insects and seeds. Sexes generally alike. Nest, chiefly in tree-holes. Eggs, 4-14. Incubation generally by female.

		Plate
1789.	SULTAN TIT *Melanochlora sultanea*	72(20)
1794.	GREY TIT *Parus major*	72(18)
1798.	WHITEWINGED BLACK TIT *Parus nuchalis*	72(19)
1799.	GREENBACKED TIT *Parus monticolus*	72(17)
1800.	YELLOWBREASTED BLUE TIT or AZURE TIT *Parus cyanus flavipectus*	72(15)
1800a.	(TIEN SHAN) YELLOWBREASTED BLUE TIT *Parus cyanus tianschanicus*	72(16)
1802.	CRESTED BLACK TIT *Parus melanolophus*	72(12)
1803.	COAL TIT *Parus ater*	72(10)
1804.	BLACK TIT *Parus rufonuchalis*	72(11)
1805.	RUFOUSBELLIED CRESTED TIT *Parus rubidiventris*	72(13)
1808.	BROWN CRESTED TIT *Parus dichrous*	72(9)
1809.	YELLOWCHEEKED TIT *Parus xanthogenys*	72(22)
1812.	BLACKSPOTTED YELLOW TIT *Parus spilonotus*	72(14)
1814.	YELLOWBROWED TIT *Sylviparus modestus*	72(21)
1815.	FIRECAPPED TIT *Cephalopyrus flammiceps*	72(6)
1817.	PENDULINE TIT *Remiz pendulinus*	72(7)
EL.	Ssp. *nigricans* of 1817	72(8)
1818.	1819 REDHEADED TIT *Aegithalos concinnus*	72(4)
1820.	Ssp. *manipurensis* of 1818	72(5)
1821.	WHITECHEEKED TIT *Aegithalos leucogenys*	72(1)
1822.	WHITETHROATED TIT *Aegithalos niveogularis*	72(2)
1823.	RUFOUSFRONTED TIT *Aegithalos iouschistos*	72(3)
EL.	BLUE-GREY TIT *Parus bokharensis*	72(23)

Family SITTIDAE: Nuthatches, Creepers

Small arboreal climbing birds. Plumage blue, grey with black or brown; usually a dark line and white markings on tail present. Body compact. Bill slender, straight and notched. Tarsi short; toes long; claws laterally compressed. Wings long and pointed; tail short and truncate. Sexes nearly alike. Flight undulating. Climb tree trunks and up and around branches without using tail as support, with jerky hops or obliquely. Food, insects, nuts, seeds. Nest, in lined tree holes or crevices. Eggs, 4-14.

		Plate
1824.	EUROPEAN NUTHATCH *Sitta europaea cashmirensis*	73(4)
1826.	EUROPEAN NUTHATCH *Sitta europaea nagaensis*	73(1)
1830.	CHESTNUTBELLIED NUTHATCH *Sitta castanea*	73(3)
1832.	WHITECHEEKED NUTHATCH *Sitta leucopsis*	73(6)
1834.	WHITETAILED NUTHATCH *Sitta himalayensis*	73(2)
1836.	ROCK NUTHATCH *Sitta tephronota*	73(5)
1837.	BEAUTIFUL NUTHATCH *Sitta formosa*	73(8)
1838.	VELVETFRONTED NUTHATCH *Sitta frontalis*	73(7)
1839.	WALL CREEPER *Tichodroma muraria*	73(9)
1841.	SPOTTED GREY CREEPER *Salpornis spilonotos*	73(14)

Family CERTHIIDAE: Tree Creepers

Small arboreal birds with plumage chiefly brown to black above and white, grey or buff below. Bill slender, laterally compressed and downcurved. Toes and claws long. Wings rounded; tail longish, stiff and pointed. Sexes nearly alike. Climb tree trunks and boughs in the manner of woodpeckers, using the tail as a support. Flight undulating and not sustained. Food, chiefly tiny insects lurking in the bark. Nest, in holes or crevices in tree trunks. Eggs, 3-9.

		Plate
1842.	TREE CREEPER *Certhia familiaris*	73(10)
1847.	HIMALAYAN TREE CREEPER *Certhia himalayana*	73(11)
1849.	SIKKIM TREE CREEPER *Certhia discolor*	73(13)
1851.	NEPAL TREE CREEPER *Certhia nipalensis*	73(12)

Family MOTACILLIDAE: Pipits and Wagtails

Slender-bodied terrestrial birds. Plumage black, grey, olive, yellow or brown. Bill slender and pointed. Toes long; hind toe generally elongated. Wings pointed; tail long, constantly 'wagged' vertically. Food, chiefly tiny insects and other invertebrates, living near or in the grass. Many species migratory. Nest, open, cup-shaped, generally on the ground. Eggs, 3-7.

		Plate
1852.	INDIAN TREE PIPIT *Anthus hodgsoni*	95(8)
1854.	TREE PIPIT *Anthus trivialis*	95(9)
1856.	MEADOW PIPIT *Anthus pratensis*	95(5)
1857.	Migrant ssp. *richardi* of 1858	95(1)
1858.	PADDYFIELD PIPIT *Anthus novaeseelandiae*	95(2)

	Plate
1861. TAWNY PIPIT *Anthus campestris*	95(3)
1864. REDTHROATED PIPIT *Anthus cervinus*	95(6)
1865. VINACEOUSBREASTED PIPIT *Anthus roseatus*	95(7)
1868. BROWN ROCK PIPIT *Anthus similis*	95(4)
1870. NILGIRI PIPIT *Anthus nilghiriensis*	95(11)
1871. WATER PIPIT or ALPINE PIPIT *Anthus spinoletta*	95(10)
1873. UPLAND PIPIT *Anthus sylvanus*	95(12)
1874. FOREST WAGTAIL *Motacilla indica*	95(15)
1876. YELLOW WAGTAIL *Motacilla flava*	95(16)
1878. Ssp. *melanogrisea* of 1876	95(17)
1883. YELLOWHEADED WAGTAIL *Motacilla citreola*	95(20)
1884. GREY WAGTAIL *Motacilla cinerea*	95(19)
1885, 1886. PIED or WHITE WAGTAIL *Motacilla alba*	95(13,14)
1891. LARGE PIED WAGTAIL *Motacilla maderaspatensis*	95(18)

Family DICAEIDAE: Flowerpeckers

Small, restless, active arboreal birds. In some species both sexes dull-coloured; in others male bright plumaged. Bill shortish, thin and curved or stout. Legs short. Wings long; tail short. Constantly flit among tree-tops, uttering sharp call notes. Food, insects, nectar and berries particularly mistletoe. Some species important flower pollinators. Nest, normally purse-shaped, of soft fibres etc. pendant from a branch. Eggs, 1-5.

	Plate
1892. THICKBILLED FLOWERPECKER *Dicaeum agile*	98(1)
1895. YELLOWVENTED FLOWERPECKER *Dicaeum chrysorrheum*	98(5)
1896. YELLOWBELLIED FLOWERPECKER *Dicaeum melanoxanthum*	98(2)
1897. LEGGE'S FLOWERPECKER *Dicaeum vincens*	98(6)
1898. ORANGEBELLIED FLOWERPECKER *Dicaeum trigonostigma*	98(7)
1899. TICKELL'S FLOWERPECKER *Dicaeum erythrorhynchos*	98(3)
1902. PLAINCOLOURED FLOWERPECKER *Dicaeum concolor*	98(4)
1904. SCARLETBACKED FLOWERPECKER *Dicaeum cruentatum*	98(10)
1905. FIREBREASTED FLOWERPECKER *Dicaeum ignipectus*	98(8)

Family NECTARINIIDAE: Sunbirds, Spiderhunters

Small, active arboreal 'flower-birds'. Sexes dimorphic. Males generally glistening metallic red, yellow, green, blue, purple and black in combinations; females generally dull coloured. Bill long, slender, curved and very finely serrate. Tongue long and tubular (suctorial). Legs short and strong; claws sharp. Wings short and rounded; tail short and truncate to long and pointed. Food, flower-nectar and insects. Nest, typically purse-shaped, pendant, often with a porch overhanging the lateral entrance. Eggs, 1-3.

	Plate
1906. RUBYCHEEK *Anthreptes singalensis*	97(3)
1907, 1908. PURPLERUMPED SUNBIRD *Nectarinia zeylonica*	97(4)
1909. SMALL SUNBIRD *Nectarinia minima*	97(7)

		Plate
1910.	VAN HASSELT'S SUNBIRD *Nectarinia sperata*	97(6)
1911,	1912. LOTEN'S SUNBIRD *Nectarinia lotenia*	97(9)
1913.	OLIVEBACKED SUNBIRD *Nectarinia jugularis*	97(5)
1917.	PURPLE SUNBIRD *Nectarinia asiatica*	97(10)
1919.	MRS GOULD'S SUNBIRD *Aethopyga gouldiae*	97(8)
1923.	NEPAL YELLOWBACKED SUNBIRD *Aethopyga nipalensis*	97(11)
1925.	BLACKBREASTED SUNBIRD *Aethopyga saturata*	97(12)
1927, 1929a.	YELLOWBACKED SUNBIRD *Aethopyga siparaja*	97(13)
1930.	FIRETAILED SUNBIRD *Aethopyga ignicauda*	97(16)
1931.	LITTLE SPIDERHUNTER *Arachnothera longirostris*	97(14)
1932.	STREAKED SPIDERHUNTER *Arachnothera magna*	97(15)

Family ZOSTEROPIDAE: White-eyes

Gregarious, active, restless arboreal birds. Plumage typically olive-green above and yellow and white below. Conspicuous white eye-ring or 'spectacle' present. Bill slender, pointed and slightly curved. Legs short and strong. Wings pointed; tail truncate. Sexes alike. Food, insects, fruit and nectar. Voice, plaintive jingling notes; song, sometimes a far-carrying warble. Nest, a deep cup cemented with cobweb, placed in the fork of a twig. Eggs, 2-4.

		Plate
1933.	WHITE-EYE *Zosterops palpebrosa*	97(1)
1937.	CEYLON WHITE-EYE *Zosterops ceylonensis*	97(2)

Family PLOCEIDAE: Weaver Birds

Small sized gregarious, arboreal birds, mainly vegetarian. Plumage, from grey and brown, variously patterned, to mainly shades of red, yellow, purple, blue, green or black in variegated combinations; much white in snow finches. Bill thick and sharp-pointed to massive; not very long. Legs short and stout. Wings short and rounded to long and pointed. Flight strong and direct in most species. Food, chiefly grains, grass seeds, berries, etc., also insects. Nest, variously constructed, of grass etc., including some very elaborate and skilfully woven and pendant. Eggs, 2-8.

Subfamily PASSERINAE: House and Rock Sparrows

		Plate
1938.	HOUSE SPARROW *Passer domesticus*	99(2)
1940.	SPANISH SPARROW *Passer hispaniolensis*	99(3)
1942.	TREE SPARROW *Passer montanus*	99(7)
1945.	SIND JUNGLE SPARROW *Passer pyrrhonotus*	99(5)
1946.	CINNAMON TREE SPARROW *Passer rutilans*	99(6)
1947a.	SCRUB SPARROW or DEAD SEA SPARROW *Passer moabiticus*	99(4)
1949.	YELLOWTHROATED SPARROW *Petronia xanthocollis*	99(9)
1950.	ROCK SPARROW *Petronia petronia*	99(8)
1951.	PALLAS'S SNOW FINCH *Montifringilla nivalis*	100(13)
1952.	TIBET SNOW FINCH *Montifringilla adamsi*	100(12)

		Plate
1953.	MANDELLI'S SNOW FINCH *Montifringilla taczanowskii*	100(14)
1954.	REDNECKED SNOW FINCH *Montifringilla ruficollis*	100(16)
1955.	BLANFORD'S SNOW FINCH *Montifringilla blanfordi*	100(10)
1956.	PERE DAVID'S SNOW FINCH *Montifringilla davidiana*	100(11)
EL.	SAXAUL SPARROW *Passer ammodendri*	99(1)
EL.	BARTAILED SNOW FINCH *Montifringilla theresae*	100(18)

Subfamily PLOCEINAE: Weaver Birds, Bayas

		Plate
1957.	BAYA *Ploceus philippinus*	99(13)
1959.	Ssp. *burmanicus* of 1957	99(14)
1960.	FINN'S BAYA *Ploceus megarhynchus*	99(11)
1961.	BLACKTHROATED WEAVER BIRD *Ploceus benghalensis*	99(10)
1962.	STREAKED WEAVER BIRD *Ploceus manyar*	99(12)

Subfamily ESTRILDINAE: Avadavat, Munias

		Plate
1964.	RED MUNIA or AVADAVAT *Estrilda amandava*	98(9)
1965.	GREEN MUNIA *Estrilda formosa*	98(13)
1966.	COMMON SILVERBILL or WHITETHROATED MUNIA *Lonchura malabarica*	98(17)
1968.	WHITEBACKED MUNIA *Lonchura striata*	98(19)
1971.	Ssp. *jerdoni* of 1973	98(15)
1973.	RUFOUSBELLIED MUNIA *Lonchura kelaarti*	98(16)
1974.	NUTMEG MANNIKIN or SPOTTED MUNIA *Lonchura punctulata*	98(14)
1977.	Ssp. *atricapilla* of 1978	98(11)
1978.	BLACKHEADED MUNIA *Lonchura malacca*	98(12)
1978a.	JAVA SPARROW *Padda oryzivora*	98(18)

Family FRINGILLIDAE: Finches

Seed-eating arboreal or terrestrial birds. Plumage, from cryptically patterned grey and brown to bold combinations of yellow, red, purple, blue, green, black, and white. Bill typically short, conical, thick to very massive. Wings short and rounded to long and pointed. Sexes usually dimorphic. Nest, cup-shaped and compact; of grass, wool, tow, etc. Eggs, 2-6.

Subfamily FRINGILLINAE: Chaffinches

		Plate
1979.	CHAFFINCH *Fringila coelebs*	101(1)
1980.	BRAMBLING *Fringilla montifringilla*	101(2)

Subfamily CARDUELINAE: Rosefinches and Allies

		Plate
1981.	HAWFINCH *Coccothraustes coccothraustes*	104(8)
1982.	BLACK-AND-WHITE GROSBEAK *Coccothraustes icterioides*	104(9)
1983.	ALLIED GROSBEAK *Coccothraustes affinis*	104(10)
1985.	WHITEWINGED GROSBEAK *Coccothraustes carnipes*	104(13)
1986.	SPOTTEDWINGED GROSBEAK *Coccothraustes melanozanthos*	104(11)
1989.	GOLDFINCH *Carduelis carduelis*	101(10)
1990.	HIMALAYAN GREENFINCH *Carduelis spinoides*	101(6)
1993.	TIBETAN SISKIN *Serinus thibetanus*	101(4)
1994.	LINNET *Acanthis cannabina*	101(8)
1995, 1996.	TWITE *Acanthis flavirostris*	101(9)
1997.	REDBROWED FINCH *Callacanthis burtoni*	104(3)
1998.	GOLDFRONTED FINCH *Serinus pusillus*	101(3)
2000.	HODGSON'S MOUNTAIN FINCH *Leucosticte nemoricola*	100(15)
2003.	BRANDT'S MOUNTAIN FINCH *Leucosticte brandti*	100(17)
2006.	TRUMPETER BULLFINCH *Carpodacus githagineus*	101(13)
2007.	MONGOLIAN TRUMPETER BULLFINCH *Carpodacus mongolicus*	101(14)
2008.	LICHTENSTEIN'S DESERT FINCH *Rhodospiza obsoleta*	101(11)
2009.	CRIMSONWINGED DESERT FINCH *Callacanthis sanguinea*	101(12)
2013.	COMMON ROSEFINCH or SCARLET GROSBEAK *Carpodacus erythrinus*	102(3)
2015.	NEPAL ROSEFINCH *Carpodacus nipalensis*	102(2)
2016.	BLANFORD'S ROSEFINCH *Carpodacus rubescens*	102(1)
2017.	PINKBROWED ROSEFINCH *Carpodacus rhodochrous*	102(4)
2017a.	VINACEOUS ROSEFINCH *Carpodacus vinaceus*	102(7)
2018.	REDMANTLED ROSEFINCH *Carpodacus rhodochlamys*	102(13)
2019.	SPOTTEDWINGED ROSEFINCH *Carpodacus rhodopeplus*	102(8)
2020.	Ssp. *blythi* of 2021	102(11)
2021.	WHITEBROWED ROSEFINCH *Carpodacus thura*	102(10)
2023.	BEAUTIFUL ROSEFINCH *Carpodacus pulcherrimus*	102(5)
2025.	LARGE ROSEFINCH *Carpodacus edwardsii*	102(6)
2026.	THREEBANDED ROSEFINCH *Carpodacus trifasciatus*	102(9)
2027.	GREAT ROSEFINCH *Carpodacus rubicilla*	102(15)
2028.	EASTERN GREAT ROSEFINCH *Carpodacus rubicilloides*	102(12)
2031.	REDBREASTED ROSEFINCH *Carpodacus puniceus*	102(16)
2032.	CROSSBILL *Loxia curvirostra*	102(17)
2033.	REDHEADED ROSEFINCH *Propyrrhula subhimachala*	102(14)
2034.	SCARLET FINCH *Haematospiza sipahi*	104(2)
2035.	GOLDHEADED BLACK FINCH *Pyrrhoplectes epauletta*	104(12)
2036.	BROWN BULLFINCH *Pyrrhula nipalensis*	104(4)
2038.	BEAVAN'S BULLFINCH *Pyrrhula erythaca*	104(7)
2039.	REDHEADED BULLFINCH *Pyrrhula erythrocephala*	104(6)
2040.	ORANGE BULLFINCH *Pyrrhula aurantiaca*	104(5)
EL.	EUROPEAN GREENFINCH *Carduelis chloris*	101(5)
EL.	BLACKHEADED GREENFINCH *Carduelis ambigua*	101(7)
EL.	PALE ROSEFINCH *Carpodacus synoicus*	101(15)

Family EMBERIZIDAE: Buntings

Small, chiefly terrestrial birds with the plumage brown, grey or olive with combinations of black, white, yellow, green or red. Bill short, conical and finely pointed; edges of mandibles showing a gap midway between gape and tip. Wings short and rounded to long and pointed; tail long, graduated or forked. Legs of medium length; feet large. Food, mainly seeds. Nest, cup-shaped or domed, made of grass, roots, moss, etc., built on the ground or in a bush or tree. Eggs, 1-7.

		Plate
2041.	CORN BUNTING *Emberiza calandra*	103(1)
2042.	PINE BUNTING *Emberiza leucocephalos*	103(3)
2043.	BLACKHEADED BUNTING *Emberiza melanocephala*	103(12)
2044.	REDHEADED BUNTING *Emberiza bruniceps*	103(16)
2045.	CHESTNUT BUNTING *Emberiza rutila*	103(13)
2046.	YELLOWBREASTED BUNTING *Emberiza aureola*	103(11)
2047.	BLACKFACED BUNTING *Emberiza spodocephala*	103(14)
2048.	WHITECAPPED BUNTING *Emberiza stewarti*	103(5)
2049.	ORTOLAN BUNTING *Emberiza hortulana*	103(7)
2050.	GREYNECKED BUNTING *Emberiza buchanani*	103(4)
2051.	ROCK BUNTING *Emberiza cia*	103(6)
2055.	GREYHEADED BUNTING *Emberiza fucata*	103(10)
2056.	LITTLE BUNTING *Emberiza pusilla*	103(9)
2057.	STRIOLATED BUNTING *Emberiza striolata*	103(8)
2058.	REED BUNTING *Emberiza schoeniclus*	103(15)
2060.	CRESTED BUNTING *Melophus lathami*	104(14)
EL.	YELLOWHAMMER *Emberiza citrinella*	103(2)

PLATE 1

1. (1) BLACKTHROATED DIVER *Gavia arctica* V Duck±: 65 cm. Single record from Ambala, Punjab.

2. (2) REDTHROATED DIVER *Gavia stellata* V Duck±: 61 cm. Sea coast. Single record from the Makran coast of Pakistan.

3. (3) GREAT CRESTED GREBE *Podiceps cristatus* M Duck-: 50 cm. Jheels and littoral waters. Chiefly N and C Subcontinent; southernmost, Andhra Pradesh (Anantagiri).

4. (4) BLACKNECKED GREBE *Podiceps nigricollis* V Duck-: 33 cm. Lakes. Pakistan, NW India. Recorded from Uttar Pradesh, Nepal, Sikkim, Bihar, Assam and Maharashtra.

5. (4a) REDNECKED GREBE *Podiceps griseigena* V Duck-: 42 cm. Observed on Nammal Lake, Punjab Salt Range; also on Pong Dam Lake, Himachal and Nayri Reservoir, Rajkot, Gujarat.

6. (5) LITTLE GREBE *Tachybaptus ruficollis* R Pigeon±: 23 cm. Village tanks and ponds. Subcontinent. Absent Andamans.

7. (351) MASKED FINFOOT *Heliopais personata* R Duck±: 56 cm. Dense swampy forest. NE India. Bangladesh.

PLATE 1
1 W
W
1 S
3 W
W
2 S
3 S
5 W
W
4 W
W
W
5 S
4 S
6 S
W
6 S
7

PLATE 2
1
1
2
3
4
4
5
6
7
7
8
9
10
10
11
12
12
13
13
John H. Dick

PLATE 2

1. (14) WILSON'S STORM PETREL *Oceanites oceanicus* R. Bulbul±: 19 cm. Pakistan and Indian sea coast. Sri Lanka, Maldive Is.

2. (16) LEACH'S or FORKTAILED, STORM PETREL *Oceanodroma leucorhoa* V Bulbul±: 17-19 cm. Pelagic. Two records, Colombo.

3. (EL) MATSUDAIRA'S STORM PETREL *Oceanodroma matsudairae* Japan.

4. (15) DUSKYVENTED STORM PETREL *Fregetta tropica* V Bulbul±: 20 cm Pelagic. Two records, Bay of Bengal and Car Nicobar.

5. (13b) BULWER'S GADFLY PETREL *Bulweria bulwerii* V Pigeon-: 26.5-28 cm. Single record, Maldive Is.

6. (13a) JOUANIN'S GADFLY PETREL *Bulweria fallax* R. Pigeon-: 29-30 cm. Pelagic. Arabian sea, Gulf of Aden. Sri Lanka (specimen).

7. (9) WEDGETAILED SHEARWATER *Procellaria pacifica* R Duck-: 48 cm. Pelagic. Warmer parts of Indian Ocean: Bay of Bengal; Sri Lanka; Maldives.

8. (8) PINKFOOTED SHEARWATER *Procellaria carneipes* V Duck-: 50 cm. Pelagic. Kanyakumari, Tamil Nadu (specimens). Two records, Sri Lanka. Maldives (sight).

9. (EL) SHORT-TAILED SHEARWATER *Procellaria tenuirostris* W Pacific ocean.

10. (EL) STREAKED SHEARWATER *Calonestris leucomelas* W Pacific ocean.

11. (EL) WHITEFACED STORM PETREL *Pelagodroma marina* Atlantic and Australian seas.

12. (6) CAPE PETREL *Daption capensis* V Pigeon+: 36 cm. Pelagic. Single record, Sri Lanka.

13. (11) AUDUBON'S SHEARWATER *Procellaria lherminieri* R Pigeon-: 30 cm. Pelagic. *Breeding:* Maldive Is. Coast of NW India; taken off Bombay harbour (Maharashtra).

PLATE 3

1. (21) SPOTTEDBILLED PELICAN *Pelecanus philippensis* RM Vulture+: 152 cm. Large lakes, reservoirs and rivers. Subcontinent; Sri Lanka; Nicobars (accidental).

2. (20) ROSY or WHITE PELICAN *Pelecanus onocrotalus* RM Vulture+: 183 cm. Large bodies of water. *Breeding:* Kutch. *Winter:* Pakistan, N and C India. Two Russian ringed birds recovered in Gujarat and Madhya Pradesh.

3. (22) DALMATIAN PELICAN *Pelecanus philippensis crispus* M Vulture+: 183 cm. N India, Pakistan, Bangladesh.

4. (32) LEAST FRIGATE BIRD *Fregata ariel* R Kite+: 80 cm. Pelagic. Maldive Is. *(breeding)*. Stragglers from Bombay, Kerala and Sri Lanka.

5. (31) LESSER FRIGATE BIRD *Fregata minor* V Kite+: 87-102 cm. Pelagic. Recorded, Andhra Pradesh, Kerala, Tamil Nadu and Sri Lanka.

6. (74) LESSER FLAMINGO *Phoeniconaias minor* RM Duck+: 90-105 cm. Brackish lakes. NW India; Pakistan. Straggler elsewhere. *Breeding:* Kutch.

7. (73) FLAMINGO *Phoenicopterus roseus* RM Vulture+: 140 cm. Brackish lakes and lagoons, sea coast, estuaries, mudflats. Subcontinent; Sri Lanka. *Breeding:* Great Rann of Kutch.

8. (86) WHOOPER SWAN *Cygnus cygnus* V Vulture+: 152 cm. Large rivers. Less than a dozen records from Pakistan, NW India, Nepal.

9. (87) MUTE SWAN *Cygnus olor* V Vulture+: 152 cm. Jheels. A dozen records in all from Pakistan, NW India.

10. (84) BEWICK'S SWAN *Cygnus cygnus bewickii* V Vulture+: 122 cm. Lakes. Three records from Pakistan, one from Delhi, one of the eastern subspecies *jankowskii* from Kutch.

1
2
3
1
2
3
Flying silhouette of Frigate Birds
5 ♂
4 ♂
4 ♀
5 ♀
4 imm
5 imm
6
7
8
8
7
6
10
9

PLATE 4

PLATE 4

1. (26) CORMORANT *Phalacrocorax carbo* RM Duck+: 80 cm. Inland waters and tidal lagoons. Occasionally high-altitude Himalayan lakes up to 3450 m. Subcontinent; Sri Lanka.

2. (27) INDIAN SHAG *Phalacrocorax fuscicollis* RM Duck+: 63 cm. Jheels, rivers, reservoirs and estuaries. Subcontinent and Sri Lanka. Not Himalayas.

3. (28) LITTLE CORMORANT *Phalacrocorax niger* RM Crow+: 51 cm. Lakes, reservoirs, tidal creeks. Subcontinent; Sri Lanka. Absent in Himalayas and N Pakistan.

4. (28a) PYGMY CORMORANT *Phalacrocorax pygmaeus* V Little cormorant-. Single record, Baluchistan.

5. (29) DARTER *Anhinga rufa* RM Duck+: 90 cm. Lakes and reservoirs. Subcontinent; Sri Lanka.

6. (59) BITTERN *Botaurus stellaris* M Village hen+: 71 cm. Swampy reed-beds. Subcontinent, Sri Lanka.

7. (49) LITTLE EGRET *Egretta garzetta* R Village hen+: 63 cm. Marshes, flooded paddyfields, etc. Subcontinent: Also Lakshadweep, Andamans and Nicobars; Sri Lanka; Maldives.

8. (53) MALAY or TIGER BITTERN *Gorsachius melanolophus* RM Village hen+: 51 cm. Marshy pockets and streams in tropical evergreen forest. Patchy in Subcontinent. Sri Lanka, Nicobar Is.

9. (42) POND HERON or PADDYBIRD *Ardeola grayii* R Village hen+: 46 cm. Marshes, streams, paddyfields, ponds, etc. Subcontinent; Sri Lanka.

10. (52) NIGHT HERON *Nycticorax nycticorax* R Village hen+: 58 cm. Inland waters, estuaries, coastal lagoons and backwaters. Subcontinent including Andamans and Nicobars; Sri Lanka.

1
1 imm
2
2 imm
3 imm ↑
3 imm ↓
3
4
4 imm
5 ○
5 ● ↓
5 ● ↓
5 ● ↑
5 imm ↑
6 ↑
6 ↓
6 imm ↑
John H. Dick

PLATE 5

1. (18) REDTAILED TROPIC-BIRD *Phaethon rubricauda* R Tern±: 36 cm + tail ribbons 48 cm. Pelagic. Tropical W Indian Ocean; also Nicobars, Stragglers Bay of Bengal.

2. (17) REDBILLED or SHORT-TAILED TROPIC-BIRD *Phaethon aethereus* R Tern±: 40 cm + tail ribbons 30 cm. Pelagic. N Indian Ocean. Recorded off Pakistan coast, specimens from west coast (Bombay, Kerala), Lakshadweep, Andamans, Sri Lanka.

3. (23) MASKED BOOBY *Sula dactylatra* M Duck+: 80 cm. Pelagic. Pakistan. Occasionally western seaboard and Sri Lanka. Maldives (sight).

4. (19) LONGTAILED TROPIC-BIRD *Phaethon lepturus* R Tern±: 38 cm + tail ribbons 45 cm. Pelagic. Maldive Is. (breeding). Straggler to NE India (Cachar), Andamans, Car Nicobar and Sri Lanka.

5. (24) REDFOOTED BOOBY *Sula sula* M Duck-: 41 cm. Pelagic. Lakshadweep Is., Bay of Bengal; Sri Lanka.

6. (25) BROWN BOOBY *Sula leucogaster* M Duck+: 76 cm. Pelagic. Gujarat, Malabar (Kerala) and Sri Lanka coasts, Lakshadweep Is. and Bay of Bengal.

PLATE 6

1. (38) LITTLE GREEN HERON *Ardeola striatus* R Village hen±: 46 cm. Inland waters and swamps. Subcontinent, Andamans, Nicobars, Lakshadweep, Sri Lanka, Maldives.

2. (34) GIANT HERON *Ardea goliath* V Vulture+: 142-152 cm. Lakes, swamps, estuaries and tidal creeks. Isolated records from Pakistan, Madhya Pradesh, W.Bengal, Bangladesh and Sri Lanka.

3. (33) GREAT WHITEBELLIED HERON *Ardea insignis* R Vulture+: 127 cm. Swamps, lakes and rivers. NE India, Bhutan, Nepal, Bihar north of Ganges, Bangladesh and N Burma. One December record from Tamil Nadu.

4. (36) GREY HERON *Ardea cinerea* RM Vulture+: 98 cm; standing 75 cm. Swamps, estuaries and rocky offshore islets. Subcontinent, Andamans, Nicobars, Sri Lanka; Maldives.

5. (37) PURPLE HERON *Ardea purpurea* RM Vulture+: 97 cm; standing 70 cm. Jheels, reedy lakes and rivers. Subcontinent; also Andamans, Nicobars, Sri Lanka.

6. (EL) SQUACCO HERON *Ardeola ralloides* W Asia, Africa.

7. (43) CHINESE POND HERON *Ardeola bacchus* RM Village hen±: 52 cm. Inland waters. NE India; Bangladesh to Manipur; Andaman Is.

8. (42) POND HERON or PADDYBIRD *Ardeola grayii*
For distribution see Plate 4

9. (56) CHESTNUT BITTERN *Ixobrychus cinnamomeus* RM Village hen-: 38 cm. Reed-beds, flooded paddyfields, etc. Rarely coastal backwaters. Subcontinent, Andamans, Nicobars, Sri Lanka, Maldives.

10. (57) YELLOW BITTERN *Ixobrychus sinensis* RM Village hen-: 38 cm. Reed-beds, standing paddy, swamps, etc. Subcontinent, Andamans, Nicobars, Sri Lanka.

11. (55) LITTLE BITTERN *Ixobrychus minutus* RM Village hen-: 36 cm. Marshes with dense reed-beds. Pakistan, N India, Nepal and east to Assam. Kashmir up to 1800 m. Maharashtra (specimen).

12. (58) BLACK BITTERN *Ixobrychus flavicollis* RM Village hen±: 58 cm. Reedy marshes. Subcontinent up to 1200 m and Sri Lanka. Maldives (winter). Absent in Andaman and Nicobar Is.

13. (59) BITTERN *Botaurus stellaris*
For distribution see Plate 4

14. (52) NIGHT HERON *Nycticorax nycticorax*
For distribution see Plate 4

15. (53) MALAY or TIGER BITTERN *Gorsachius melanolophus*
For distribution see Plate 4

PLATE 6

1
2
3
4
5
6 br −
6 br +
7
8 br −
8 br +
9
10 imm
10 ♂
11
12
13
14 imm
14
15
John H. Dick

1
2
3
4 ●
4 o
5 br ♀
5 br ♂
Little Egret
in flight
Cattle
Egret
in flight
Heron
Crane
6 imm
6
7
8 →
9
10
11

PLATE 7

1. (46) LARGE EGRET *Ardea alba* RM Vulture±: 91 cm. Marshes, rivers, etc. Subcontinent; Sri Lanka. *Winter:* Andamans, Nicobars (?), Maldives. Straggler to Pakistan and Uttar Pradesh.

2. (47) SMALLER EGRET *Egretta intermedia* RM Village hen+: 80 cm. Marshes, estuaries, swamps. Subcontinent; Andamans, Nicobars, Sri Lanka.

3. (49) LITTLE EGRET *Egretta garzetta*
For distribution see Plate 4

4. (50) INDIAN REEF HERON *Egretta gularis* RM Village hen±: 63 cm. Rocky seashores, tidal lagoons and mudflats, mangrove swamps. Pakistan, W India, Sri Lanka, Lakshadweep Is. Rare on east coast.

5. (44) CATTLE EGRET *Bubulcus ibis* RM Village hen±: 51 cm. Usually associated with grazing cattle, not necessarily near water. Subcontinent, Andamans, Nicobars, Lakshadweep, Sri Lanka, Maldives.

6. (325) SIBERIAN CRANE *Grus leucogeranus* M Vulture+: standing 140 cm. Jheels and marshes. N Pakistan, Rajasthan, Uttar Pradesh. Has strayed to Sind, Bihar, Madhya Pradesh and Maharashtra.

7. (323) SARUS CRANE *Grus antigone* R Vulture+: standing 156 cm. Marshes and cultivation. Pakistan and N India, south to Mhow (Madhya Pradesh) and Godavari delta; Assam and Manipur.

8. (326) DEMOISELLE CRANE *Anthropoides virgo* M Vulture-: standing 76 cm. Winter crops, paddy stubbles, margins of jheels and tanks. Pakistan, N India to Assam, Bangladesh south to Andhra Pradesh and Karnataka.

9. (321) BLACKNECKED CRANE *Grus nigricollis* RM Vulture+: standing 156 cm. Damp fields and marshland. *Breeding:* Ladakh. *Winter:* Bhutan (2000 m and above), Arunachal Pradesh. Nepal (solo, July).

10. (320) COMMON CRANE *Grus grus* M Vulture±: standing 140 cm. Cultivated plains, jheels and sandy river-beds. Pakistan and N India, east to Bengal, south to Maharashtra and N Andhra.

11. (322) HOODED CRANE *Grus monacha* V Vulture±: standing 90 cm. Marshes. Single record from N Cachar. Apparently winter visitor in very small numbers to Manipur and Assam.

Flight silhouettes
Heron
Crane

PLATE 8

1. (66) BLACKNECKED STORK *Ephippiorhynchus asiaticus* R Vulture+: standing 135 cm. Marshes and large rivers. Subcontinent. Iris male brown, female yellow. Sri Lanka (endangered).

2. (60) PAINTED STORK *Mycteria leucocephala* RM Vulture± standing 93 cm. Large marshes. Subcontinent; Sri Lanka.

3. (63) WHITE STORK *Ciconia ciconia* M Vulture+: standing 106 cm. Marshes, moist grassland. Pakistan, NW India, Gangetic Plain, east to Bangladesh and Assam, south decreasingly through the Peninsula.

4. (65) BLACK STORK *Ciconia nigra* M Vulture+: standing 106 cm. Marshes and near rivers. Pakistan and N India, south to Deccan. Single records from Andhra, Kerala and Sri Lanka.

5. (61) OPENBILL STORK *Anastomus oscitans* R Duck+: 81 cm, standing 68 cm. Lakes and marshes. Subcontinent; Sri Lanka.

6. (62) WHITENECKED STORK *Ciconia episcopus* R Vulture+: standing 106 cm. Flooded grasslands, irrigated fields, marshes in forest, etc. Subcontinent; Sri Lanka.

7. (70) BLACK IBIS *Pseudibis papillosa* R Village hen±: 68 cm. River banks, stubble fields and jheel margins. Pakistan, and India east through Bangladesh to Burma, south to Deccan and Karnataka.

8. (72) SPOONBILL *Platalea leucorodia* RM Duck±: standing 60 cm. Marshes and rivers. Subcontinent; Sri Lanka.

9. (71) GLOSSY IBIS *Plegadis falcinellus* RM Village hen±: 52 cm. Marshes and river banks. Pakistan, and India east to Bangladesh south to Deccan. Sri Lanka, Maldives.

10. (67) ADJUTANT *Leptoptilos dubius* RM and nomadic. Vulture+: standing 120-150 cm. Marshes. N India west to Pakistan (Sind) in SW monsoon, south to Deccan. Bangladesh (sporadic).

11. (69) WHITE IBIS *Threskiornis aethiopica* R Village hen±: 75 cm. Rivers, marshes and inundated land. Subcontinent; Sri Lanka.

12. (68) LESSER ADJUTANT *Leptoptilos javanicus* RM Vulture+: standing 110-120 cm. Swamps and flooded land. N India, Bangladesh and peninsula (sporadic). Sri Lanka (declining).

PLATE 8
1
2
3
4
5
6
7
8
9
10
11
10
1
12
John H. Dick

1 ♂ S
♀
2 ♂ S
♀
3
4 ♂ S
♀
5
6 ♂ S
♀
7 ♂ S
♀
8
9 ♂ S
♀
10 ♂ S
♀
11 ♂ S
♀
12 ♂ S
♀
John H. Dick

PLATE 9

1. (100) MALLARD *Anas platyrhynchos* RM Duck±: 61 cm. Shallow jheels and marshes. Pakistan, N India, east through Bangladesh south to N Maharashtra.

2. (93) PINTAIL *Anas acuta* M Duck-: 56-74 cm. Reedy jheels, brackish lagoons and estuaries. Subcontinent; Sri Lanka; Maldives.

3. (97) SPOTBILL DUCK *Anas poecilorhyncha* RM Duck±: 61 cm. Jheels and shallow reservoirs. Subcontinent: south to Karnataka. Up to 1800 m in Kashmir. Andamans (once); Sri Lanka (occasional).

4. (103) WIGEON *Anas penelope* M Duck-: 49 cm. Reedy marshes. Pakistan and N India east through Bangladesh, south to Orissa, Tamil Nadu; Sri Lanka.

5. (96) GREY TEAL *Anas gibberifrons* R Half-grown Duck+: 43 cm. Swamps and tidal creeks. Andaman Islands.

6. (105) SHOVELLER *Anas clypeata* M Duck-: 51 cm. All types of inland waters. Subcontinent; Sri Lanka; Maldives.

7. (95) BAIKAL TEAL *Anas formosa* V Half-grown Duck+: 40 cm. Pakistan east to Assam and Manipur (a score of records in all).

8. (92) MARBLED TEAL *Marmaronetta angustirostris* RM Duck-: 48 cm. Reedy jheels. *Breeding:* Pakistan. *Winter:* N India east to Assam south to Pune.

9. (94) COMMON TEAL *Anas crecca* M Half-grown Duck±: 38 cm. Jheels and marshes. Subcontinent, Andamans, Nicobars, Sri Lanka, Maldives.

10. (104) GARGANEY *Anas querquedula* M Duck-: 41 cm. Marshes, reservoirs and lakes. Subcontinent; Sri Lanka.

11. (102) FALCATED TEAL *Anas falcata* V Duck-: 51 cm. Jheels. Pakistan, and NE India. One record each from Gujarat and Nepal.

12. (101) GADWALL *Anas strepera* M Duck-: 51 cm. Reedy marshes. Subcontinent; rare in S India. Sri Lanka (once).

PLATE 10

Ducks in flight

1. (100) MALLARD *Anas platyrhynchos*
   For distribution see Plate 9

2. (93) PINTAIL *Anas acuta*
   For distribution see Plate 9

3. (97) SPOTBILL DUCK *Anas poecilorhyncha*
   For distribution see Plate 9

4. (103) WIGEON *Anas penelope*
   For distribution see Plate 9

5. (96) GREY TEAL *Anas gibberifrons*
   For distribution see Plate 9

6. (105) SHOVELLER *Anas clypeata*
   For distribution see Plate 9

7. (95) BAIKAL TEAL *Anas formosa*
   For distribution see Plate 9

8. (92) MARBLED TEAL *Marmaronetta angustirostris*
   For distribution see Plate 9

9. (94) COMMON TEAL *Anas crecca*
   For distribution see Plate 9

10. (104) GARGANEY *Anas querquedula*
    For distribution see Plate 9

11. (102) FALCATED TEAL *Anas falcata*
    For distribution see Plate 9

12. (101) GADWALL *Anas strepera*
    For distribution see Plate 9

PLATE 10
1 ♂ S
♀
2 ♂ S
♀
3
5
4 ♂ S
♀
6 ♂ S
♀
8
7 ♂ S
♀
9 ♂ S
♀
10 ♂ S
♀
11 ♂ S
♀
12 ♂ S
♀
John H. Dick

1 ♂
2 ♂
3
4
5 ♂ S
♀
6 ♂ S
♀
7 ♂ S
♀
8 ♂ S
♀
9 ♂ S
♀
10 ♂ S
♀
11 ♂ S
♀
12 ♂ S
♀
13 ♂
John H. Dick

PLATE 11

1. (90) RUDDY SHELDUCK *Tadorna ferruginea* RM Duck+: 66 cm. Lakes and rivers. *Breeding:* Ladakh. Subcontinent; rare in S India. Sri Lanka (occasional).

2. (91) COMMON SHELDUCK *Tadorna tadorna* M Duck±: 61 cm. Lakes and large rivers. Pakistan, India, Nepal, south to Kutch, the Deccan and Orissa east through Bangladesh to Burma.

3. (89) LARGE WHISTLING TEAL *Dendrocygna bicolor* RM Duck-: 51 cm. (Upper tail-coverts white). Reedy jheels and reservoirs. Pakistan, N India east to Manipur and Bangladesh; south to the Deccan, Andamans, Nicobars, Sri Lanka. Rare.

4. (88) LESSER WHISTLING TEAL or TREE DUCK *Dendrocygna javanica* R Duck-: 42 cm. (Upper tail-coverts chestnut). Marshes, reservoirs, reedy ponds. Subcontinent, Sri Lanka.

5. (107) REDCRESTED POCHARD *Netta rufina* M Duck-: 54 cm. Large jheels and reservoirs. Pakistan and NW India, decreasingly eastward to Assam, and southward to Tamil Nadu.

6. (108) COMMON POCHARD *Aythya ferina* M Duck-: 48 cm. Open jheels and reservoirs. Pakistan and NW India, decreasingly eastward to Bangladesh, Assam and associated states, and southward to Karnataka and Pondicherry.

7. (111) TUFTED DUCK *Aythya fuligula* M Duck-: 46 cm. Open jheels and reservoirs. Pakistan, Nepal and N India, east through Bangladesh to Manipur, south decreasingly to Karnataka, Pondicherry, Sri Lanka, Maldives.

8. (112) SCAUP DUCK *Aythya marila* V Duck-: 43 cm. In India on freshwater lakes; elsewhere largely a sea duck. Pakistan, N India south to Maharashtra, Madhya Pradesh, east through Bangladesh to Burma.

9. (110) BAER'S POCHARD *Aythya baeri* M Duck-: 46 cm. Jheels and reservoirs. Assam and associated states, Manipur, Bangladesh, Bihar, Bengal, Nepal, Rajasthan and Pakistan (Punjab).

10. (109) WHITE-EYED POCHARD or FERRUGINOUS DUCK *Aythya nyroca* RM Duck-: 41 cm. Jheels and reservoirs. *Breeding:* Kashmir and Ladakh. *Winter:* Pakistan and NW India, Manipur, Bangladesh, Deccan, Kerala, Maldives.

11. (119) SMEW *Mergus albellus* M Duck-: 46 cm. Clear Himalayan foothill streams. Pakistan, Nepal terai and N India east to Assam and south to Gujarat and Orissa. Rare.

12. (114) COTTON TEAL or QUACKY-DUCK *Nettapus coromandelianus* R Partridge±: 33 cm. Jheels and ponds. Subcontinent, Andamans (straggler), Sri Lanka, Maldives.

13. (106) PINKHEADED DUCK *Rhodonessa caryophyllacea* Extinct. Duck±: 60 cm. Jheels and swampy lowland grass jungles. Formerly resident from Bihar to Assam, Manipur and Orissa. Had straggled to Punjab, Maharashtra, Andhra Pradesh, Tamil Nadu. Last reliable wild record, 1935.

PLATE 12

1 (123) WHITEHEADED STIFFTAILED DUCK *Oxyura leucocephala* M Duck-: 46 cm. Large lakes. Pakistan and NW India; Rajasthan (once).

2. (118) GOLDEN EYE DUCK *Bucephala clangula* M Duck-: 46 cm. Open reaches of swift-flowing rivers. Pakistan, Nepal (up to 3000 m) and N and NE India.

3. (117) LONGTAIL or OLD SQUAW DUCK *Clangula hyemalis* V Duck-: 42 cm. Jheels and large rivers. Three records from Pakistan; one each from Kashmir, Uttar Pradesh, Nepal and Arunachal Pradesh.

4. (120, 121) GOOSANDER, COMMON MERGANSER *Mergus merganser* RM Duck+: 66 cm. Large rivers, lakes and fast-flowing streams. Pakistan and N India south to Bombay and Madhya Pradesh (occasional), and east to Assam. *Breeding*: Ladakh.

5. (115) COMB DUCK *Sarkidiornis melanotos* R Duck+: 76 cm. Reedy tanks in well-wooded plains. Pakistan and India from Nepal terai south to Karnataka and east through Bangladesh to Burma. Formerly resident in Sri Lanka, but now believed extinct.

6. (116) WHITEWINGED WOOD DUCK *Cairina scutulata* R Duck+: 81 cm. Secluded pools, and marshes in dense forest. Formerly resident in Bangladesh, Assam, Arunachal Pradesh and Manipur; now probably confined chiefly to Arunachal Pradesh.

7. (75) SIBERIAN REDBREASTED GOOSE *Branta ruficollis* V Duck±: 61 cm. Three records: Madhya Pradesh, Assam.

8. (80) LESSER WHITEFRONTED GOOSE *Anser erythropus* M Duck±: 53 cm. Yellow eye-ring. Jheels and lakes. Pakistan, Kashmir and N India east to Assam; Maharashtra.

9. (76) BEAN GOOSE *Anser fabalis* V (?) Duck+: 76 cm. One possible sighting, Arunachal Pradesh.

10. (79) WHITEFRONTED GOOSE *Anser albifrons* M Duck+: 68 cm. No eye-ring. Large lakes. Pakistan, and N India east to Assam and Manipur. Southernmost record; Orissa.

11. (81) GREYLAG GOOSE *Anser anser* M Duck+: 81 cm. Large jheels and lakes. Pakistan, N and C India to Manipur, south to Chilka lake, Orissa. Sri Lanka (once).

12. (82) BARHEADED GOOSE *Anser indicus* RM Duck+: 75 cm. Large jheels and rivers. Pakistan, N India east through Bangladesh to Burma, south to Karnataka, Andhra and Tamil Nadu. *Breeding:* Ladakh.

1 ♂ S
♀
2 ♂ S
♀
3 ♂ S
♀
5
6
4 ♂ S
♀
♀
5 ♂ S
6 ♂
8
9
7
10
11
11
12
12
John H. Dick

PLATE 13

1 ♂

2 ♂

3

4

5 ♂↑ ♀↓

6 ♂↑ ♀↓

7 ♂↑ ♀↓

8 ♂↑ ♀↓

9 ♂↑

10 ♂↑ ♀↓

11 ♂↑ ♀↓

12 ♂↑ ♀↓

13 ♂↑

John H. Dick

PLATE 13

Ducks in flight. In all cases where pairs shown, male below, female above.

1. (90) RUDDY SHELDUCK *Tadorna ferruginea*
   For distribution see Plate 11

2. (91) COMMON SHELDUCK *Tadorna tadorna*
   For distribution see Plate 11

3. (89) LARGE WHISTLING TEAL *Dendrocygna bicolor*
   For distribution see Plate 11

4. (88) LESSER WHISTLING TEAL or TREE DUCK *Dendrocygna javanica*
   For distribution see Plate 11

5. (107) REDCRESTED POCHARD *Netta rufina*
   For distribution see Plate 11

6. (108) COMMON POCHARD *Aythya ferina*
   For distribution see Plate 11

7. (111) TUFTED DUCK *Aythya fuligula*
   For distribution see Plate 11

8. (109) WHITE-EYED POCHARD *Aythya nyroca*
   For distribution see Plate 11

9. (112) SCAUP DUCK *Aythya marila*
   For distribution see Plate 11

10. (119) SMEW *Mergus albellus*
    For distribution see Plate 11

11. (120, 121) GOOSANDER or COMMON MERGANSER *Mergus merganser*
    For distribution see Plate 12

12. (114) COTTON TEAL or QUACKY-DUCK *Nettapus coromandelianus*
    For distribution see Plate 11

13. (123) WHITEHEADED STIFFTAILED DUCK *Oxyura leucocephala*
    For distribution see Plate 12

imm
imm
imm
1
2
3
imm
imm
4
5
6
7
7 imm
10
imm
8
9
12 ♂
imm
ad –
♀
o
11
13 ♂
14 ♂
imm
John H. Dick

PLATE 14

1. (173) WHITEBELLIED SEA EAGLE *Haliaeetus leucogaster* R Kite+: 66-71 cm. Sea coast, tidal creeks and estuaries. Offshore islands and peninsular seaboard from Bombay to Bangladesh; Sri Lanka, Lakshadweep, Andaman and Nicobar Is., Gujarat coast (vagrant); also Rajasthan (Bharatpur).

2. (175) GREYHEADED FISHING EAGLE *Ichthyophaga ichthyaetus* R Kite+: 74 cm. Rivers, lakes, reservoirs in well-wooded country. Gangetic Plain east of Delhi through Bangladesh to Burma, south to Kerala, Andamans, Sri Lanka.

3. (177) HIMALAYAN GREYHEADED FISHING EAGLE *Ichthyophaga nana* R Kite±: 64 cm. Clear forest streams; foothills up to 2400 m. Himalayas, Kashmir to Arunachal Pradesh; NE hill states. *Winter:* Plains of Haryana, Uttar Pradesh and Bihar.

4. (172a) WHITETAILED EAGLE *Haliaeetus albicilla* M Kite+: 69-86 cm. Essentially a sea eagle, but wanders up large rivers and to jheels and inundations. Pakistan. Also Gujarat, Rajasthan, Himachal Pradesh, Nepal and Orissa (Chilka).

5. (174) PALLAS'S FISHING EAGLE *Haliaeetus leucoryphus* RM Kite+: 76-84 cm. Large rivers, lakes and jheels. Pakistan, and N India east through Bangladesh, south to Chilka Lake, Orissa in the east and Maharashtra (Bombay) in the west.

6. (203) OSPREY *Pandion haliaetus* RM Kite-: 56 cm. Lakes, reservoirs, jheels, estuaries. Subcontinent, Andamans, Nicobars, Lakshadweep, Sri Lanka, Maldives. *Breeding:* (rarely) in the Himalayas 2000-3800 m (Ladakh, Kashmir, Garhwal, Kumaon, Arunachal Pradesh). A bird ringed in Norway recovered in Gujarat.

7. (133) PARIAH KITE *Milvus migrans govinda* R 61 cm. Chiefly urban localities. Subcontinent, Andamans (vagrant); Sri Lanka.

8. (132) BLACK KITE *Milvus migrans migrans* R Kite: 61 cm. Chiefly urban localities. Baluchistan. *Winter:* Sind; E. Ghats (Andhra).

9. (134) BLACKEARED or LARGE INDIAN KITE *Milvus migrans lineatus* RM Kite+: 66 cm. Open wooded country. Pakistan, and India south to c. 18°N. *Breeding:* Ladakh and N Kashmir; possibly througout Himalayas

10. (131) RED KITE *Milvus milvus* V Kite: 61 cm. Five sight records: Ladakh, Kutch, Gujarat, Rajasthan and Orissa.

11. (135) BRAHMINY KITE *Haliastur indus* R Kite-: 48 cm. Sea coast and inland waters. Subcontinent (except Baluchistan), Andamans; Sri Lanka.

12. (192) PIED HARRIER *Circus melanoleucos* RM Kite-: 46-49 cm. Grasslands, ricefields, margins of jheels. Chiefly E Subcontinent. *Breeding:* Assam. Madhya Pradesh (*Winter*).

13. (190) PALE HARRIER *Circus macrourus* M Kite-: 46-51 cm. Undulating country; plateaux, grassy hillsides, cultivation and semi-desert. Subcontinent, Lakshadweep, Andamans, Sri Lanka, Maldives.

14. (193) MARSH HARRIER *Circus aeruginosus* M Kite-: 54-59 cm. Marshes, flooded ricefields, up to 2000 m in the hills. Subcontinent, Andamans, Nicobars (?), Lakshadweep, Sri Lanka, Maldives.

PLATE 15

Kites and Eagles in flight (from below)

1. (131) RED KITE *Milvus milvus*
   For distribution see Plate 14

2. (134) BLACKEARED or LARGE INDIAN KITE *Milvus migrans lineatus*
   For distribution see Plate 14

3. (133) PARIAH KITE *Milvus migrans govinda*
   For distribution see Plate 14

4. (132) BLACK KITE *Milvus migrans migrans*
   For distribution see Plate 14

5. (177) HIMALAYAN GREYHEADED FISHING EAGLE *Ichthyophaga nana*
   For distribution see Plate 14

6. (173) WHITEBELLIED SEA EAGLE *Haliaeetus leucogaster*
   For distribution see Plate 14

7. (175) GREYHEADED FISHING EAGLE *Ichthyophaga ichthyaetus*
   For distribution see Plate 14

8. (172a) WHITETAILED EAGLE *Haliaeetus albicilla*
   For distribution see Plate 14

9. (174) PALLAS'S FISHING EAGLE *Haliaeetus leucoryphus*
   For distribution see Plate 14

PLATE 15

1 imm
1 ad –
1
2
2 imm
3 imm
3
4 imm
4
5
5 imm
5 imm
5
6 imm
6
7 imm
7
8 imm
8
John H. Dick

PLATE 16

1. (186) EGYPTIAN or SCAVENGER VULTURE *Neophron percnopterus* RM Kite±: 64 cm. Near towns and villages. Subcontinent up to *c.* 3600 m. Sri Lanka (straggler).

2. (188) BEARDED VULTURE or LÄMMERGEIER *Gypaetus barbatus* R. Vulture+: 122 cm. Himalayas and associated northern mountains. Pakistan, Kashmir and east to Bhutan and Arunachal Pradesh, normally 1200 to 4200 m, soaring to over 7000 m.

3. (185) INDIAN WHITEBACKED VULTURE *Gyps bengalensis* R 90 cm. Open countryside. Subcontinent, up to *c.* 2500 m. Absent in Sri Lanka.

4. (182) INDIAN LONGBILLED VULTURE *Gyps indicus* R 92 cm. Open countryside. Subcontinent. Absent in Sri Lanka.

5. (180) GRIFFON VULTURE *Gyps fulvus*[1] RM Vulture+: 110-122 cm. Semi-desert country and bare mountains. *Breeding:* Mountains of Pakistan and Kashmir, *Winter:* Rest of Pakistan, NW India, Nepal up to *c.* 3000 m, W. Bengal, Assam and Deccan (occasional).

6. (179) CINEREOUS VULTURE *Aegypius monachus* RM Vulture+: 100-110 cm. Semi-desert. Baluchistan to Gilgit; Assam; possibly elsewhere in the dry temperate zone along the Himalayas. Vagrant in Sind, Kutch, Gujarat, Madhya Pradesh, Maharashtra and Kerala (winter).

7. (178) BLACK or KING VULTURE *Sarcogyps calvus* R 84 cm. Open country, cultivation and semi-desert. Subcontinent up to *c.* 2000 m. Absent Sri Lanka.

8. (181) HIMALAYAN GRIFFON *Gyps himalayensis* R Vulture+; 122 cm. Bare high mountains. Himalayas, *c.* 600-2500 m.

[1] No. 5 is repeated because since this plate was painted the two subspecies shown in the plate have been synonymised.

PLATE 17

Kites, Vultures and Eagles in flight (from above)

1. (133) PARIAH KITE *Milvus migrans govinda*
   For distribution see Plate 14

2. (135) BRAHMINY KITE *Haliastur indus*
   For distribution see Plate 14

3. (131) RED KITE *Milvus milvus*
   For distribution see Plate 14

4. (186) EGYPTIAN or SCAVENGER VULTURE *Neophron percnopterus*
   For distribution see Plate 16

5. (164) BOOTED HAWK-EAGLE *Hieraaetus pennatus*
   For distribution see Plate 20

6. (188) BEARDED VULTURE or LÄMMERGEIER *Gypaetus barbatus*
   For distribution see Plate 16

7. (179) CINEREOUS VULTURE *Aegypius monachus*
   For distribution see Plate 16

8. (178) BLACK or KING VULTURE *Sarcogyps calvus*
   For distribution see Plate 16

9. (181) HIMALAYAN GRIFFON *Gyps himalayensis*
   For distribution see Plate 16

10. (180) GRIFFON VULTURE *Gyps fulvus*
    For distribution see Plate 16

11. (185) INDIAN WHITEBACKED VULTURE *Gyps bengalensis*
    For distribution see Plate 16

12. (182) INDIAN LONGBILLED VULTURE *Gyps indicus*
    For distribution see Plate 16

From above

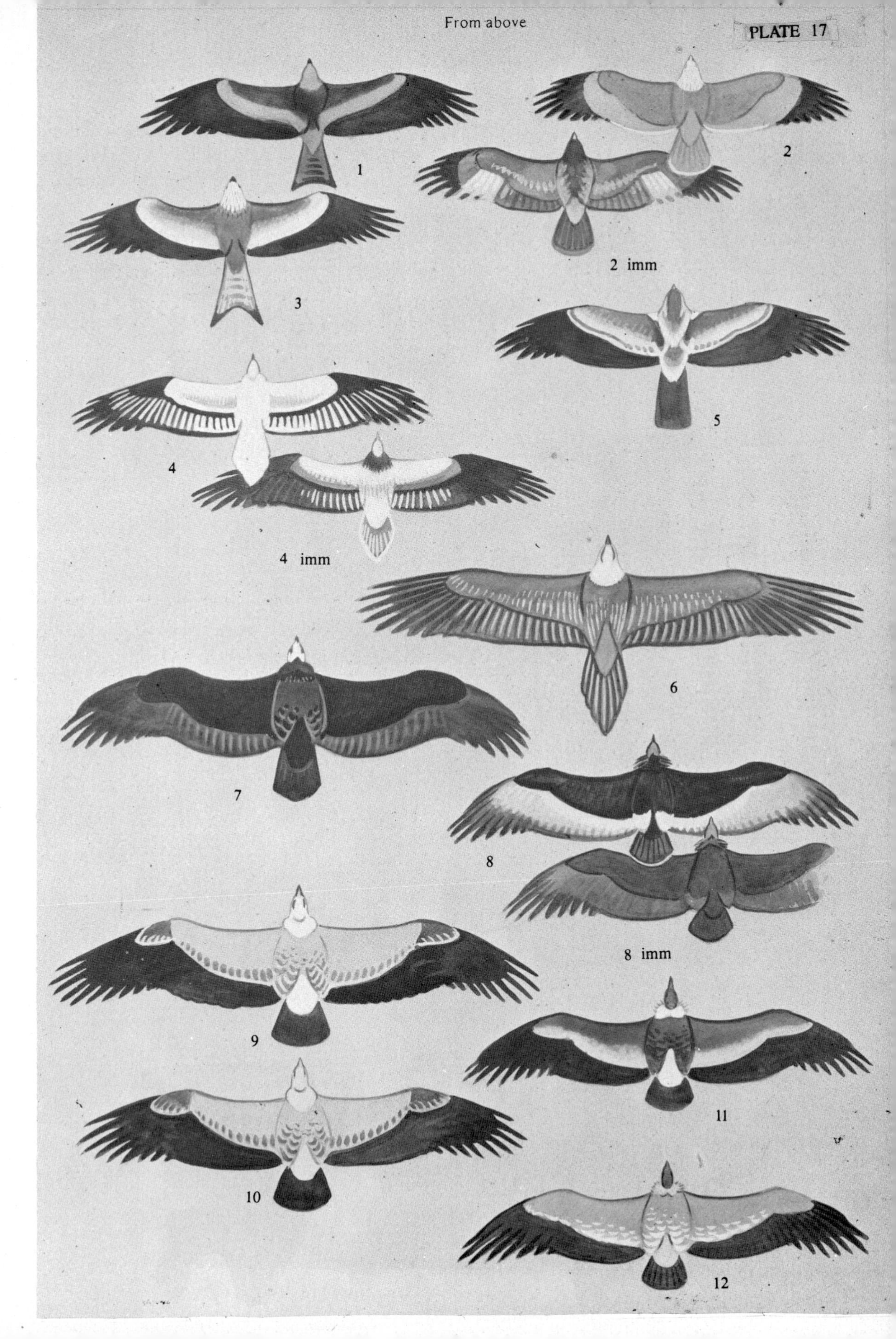

1
imm
2
imm
4
3
imm
ad -
5
imm
6
7
8
7 imm
John H. Dick

PLATE 18

Vultures in flight (from below)

1. (186) EGYPTIAN or SCAVENGER VULTURE *Neophron percnopterus*
   For distribution see Plate 16

2. (188) BEARDED VULTURE or LÄMMERGEIER *Gypaetus barbatus*
   For distribution see Plate 16

3. (179) CINEREOUS VULTURE *Aegypius monachus*
   For distribution see Plate 16

4. (178) BLACK or KING VULTURE *Sarcogyps calvus*
   For distribution see Plate 16

5. (185) INDIAN WHITEBACKED VULTURE *Gyps bengalensis*
   For distribution see Plate 16

6. (181) HIMALAYAN GRIFFON *Gyps himalayensis*
   For distribution see Plate 16

7. (182) INDIAN LONGBILLED VULTURE *Gyps indicus*
   For distribution see Plate 16

8. (180) GRIFFON VULTURE *Gyps fulvus*
   For distribution see Plate 16

PLATE 19

Harriers in flight

1. (189) HEN-HARRIER *Circus cyaneus* M Kite-: 46-54 cm. Open plains, cultivated country and foothills to *c.* 2500 m. Pakistan, and N India east to Assam, south to Maharashtra.

2. (190) PALE HARRIER *Circus macrourus*
   For distribution see Plate 14

3. (191) MONTAGU'S HARRIER *Circus pygargus* M Kite-: 46-49 cm. Swamps, grassy plains and cultivation. Subcontinent, Andamans, Lakshadweep; Sri Lanka; Maldives.

4. (192) PIED HARRIER *Circus melanoleucos*
   For distribution see Plate 14

5. (193) MARSH HARRIER *Circus aeruginosus*
   For distribution see Plate 14

6. (194) EASTERN MARSH HARRIER or STRIPED HARRIER *Circus aeruginosus spilonotus* M Kite-: 54-59 cm. Marshland. Only recorded in Manipur and Assam (Cachar). A recent sight record from Corbett, Uttar Pradesh.

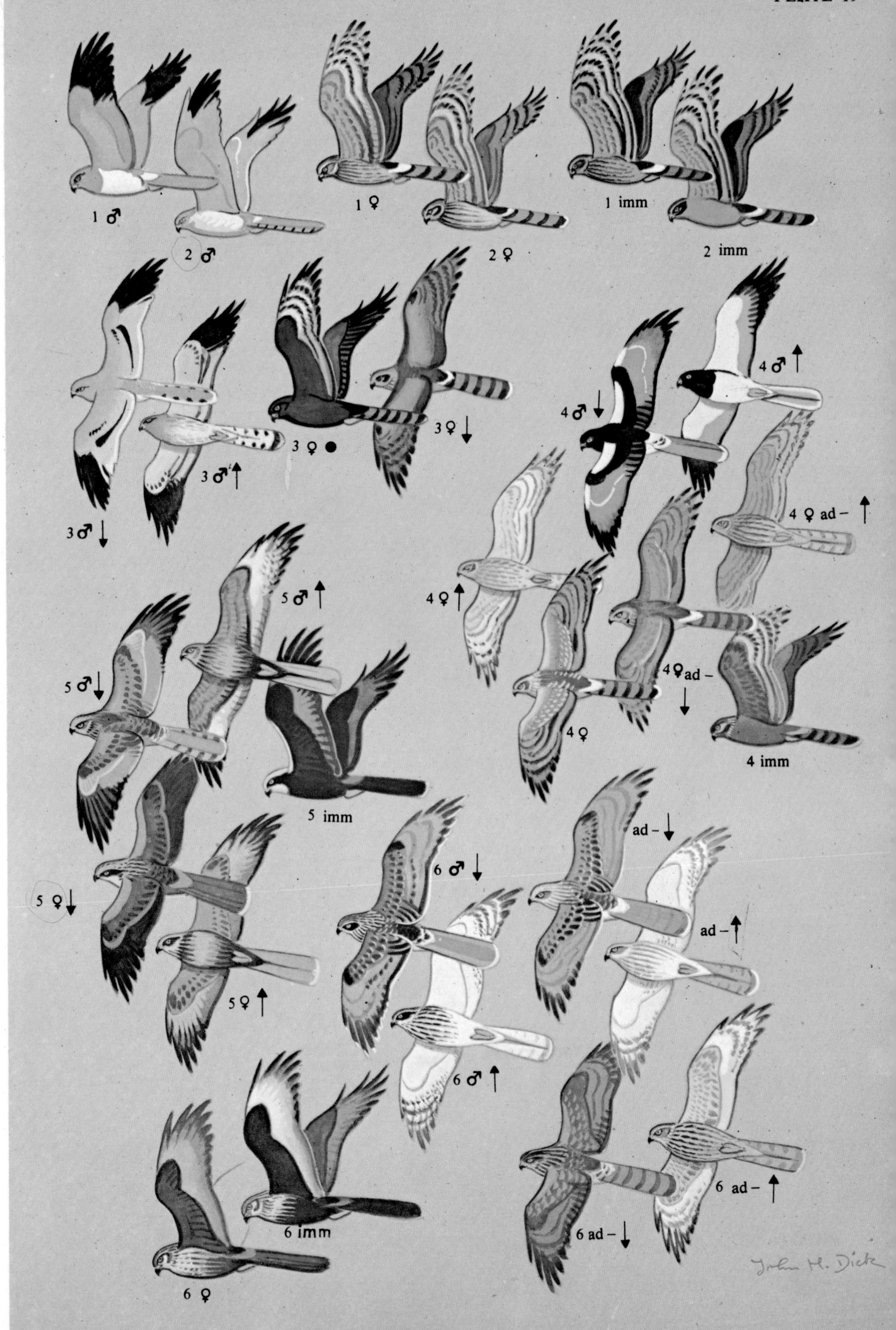
1 ♂
2 ♂
1 ♀
2 ♀
1 imm
2 imm
3 ♂ ↓
3 ♂ ↑
3 ♀ ●
3 ♀ ↓
4 ♂ ↓
4 ♂ ↑
4 ♀ ad – ↑
4 ♀ ↑
4 ♀ ad – ↓
4 ♀
4 imm
5 ♂ ↑
5 ♂ ↓
5 imm
5 ♀ ↓
5 ♀ ↑
6 ♂ ↓
6 ♂ ↑
ad – ↓
ad – ↑
6 imm
6 ♀
6 ad – ↑
6 ad – ↓
John H. Dick

1 1st year
2 imm ●
2 ○
1
2nd year
2 imm ○
2 ●
1
3 imm
3
4 ad –
6 imm
5 ○
6
6 ad –
4
5 ●
9
7
7 imm
8 imm
11
12 ad –
10
8
12
John M. Dick

PLATE 20

1. (163) BONELLI'S EAGLE *Hieraaetus fasciatus* R Kite+: 68-72 cm. Well-wooded country. Subcontinent, up to 2400 m in Himalayas; Assam (?) east of N Bengal duars. Sri Lanka (once).

2. (164) BOOTED HAWK-EAGLE *Hieraaetus pennatus* RM Kite-: 50-54 cm. Well-wooded country hill and plain, and semi-desert. W Himalayas above 1000 m. *Winter:* Subcontinent; Sri Lanka.

3. (165) RUFOUSBELLIED HAWK-EAGLE *Hieraaetus kienerii* R Kite±: 53-61 cm. Evergreen and moist-deciduous forest. Range discontinuous: W Ghats from Goa and N Karnataka south through Kerala, Tamil Nadu and Sri Lanka; Nepal and Sikkim to Arunachal Pradesh, south through Assam and Bangladesh to Burma.

4. (158) HODGSON'S HAWK-EAGLE *Spizaetus nipalensis* RM Kite+: 72 cm. Forests. Himalayas (between *c.* 600 and 2400 m), the hills south of the Brahmaputra, and Manipur. Occasional in northern plains and Peninsula in winter. Also Sri Lanka.

5. (160) CHANGEABLE HAWK-EAGLE *Spizaetus cirrhatus limnaeetus* R Kite+: 70 cm. Forests. Lower Himalayas (below *c.* 1800 m) from Garhwal to Arunachal Pradesh.

6. (161) CRESTED HAWK-EAGLE *Spizaetus cirrhatus cirrhatus* R Kite+: 72 cm. Deciduous and semi-evergreen forest. Gangetic Plain and Indian peninsula southward. Andamans. Also Sri Lanka.

7. (195) SHORT-TOED EAGLE *Circaetus gallicus* R Kite+: 63-68 cm. Cultivated plains, stony deciduous scrub, semi-desert and hills to *c.* 2300 m. Subcontinent; not east of Bangladesh or Sri Lanka.

8. (196) CRESTED SERPENT EAGLE *Spilornis cheela* R Kite+: 74 cm. Well-wooded country. Himalayan foothills and Gangetic Plain; peninsular India; Sri Lanka.

9. (200) ANDAMAN PALE SERPENT EAGLE *Spilornis cheela davisoni* R Kite-: 56 cm. Tidal creeks and mangroves. Andaman and Nicobar Is.

10. (201) NICOBAR CRESTED SERPENT EAGLE *Spilornis cheela minimus* R Kite-: 48 cm. Only in forest near rivers. Nicobar Is.[1]

11. (202) GREAT NICOBAR SERPENT EAGLE *Spilornis klossi* R Kite-: 46 cm. Sambelong or Great Nicobar I.

12. (202a) ANDAMAN DARK SERPENT EAGLE *Spilornis elgini* R Kite-: 50 cm. Inland forest clearings and sparsely wooded hillsides. Andaman Is.

[1] *S. c. malayensis* recorded from Great Nicobar.

PLATE 21

Raptors in flight (from below)

1. (203) OSPREY *Pandion haliaetus*
   For distribution see Plate 14

2. (163) BONELLI'S EAGLE *Hieraaetus fasciatus*
   For distribution see Plate 20

3. (195) SHORT-TOED EAGLE *Circaetus gallicus*
   For distribution see Plate 20

4. (196) CRESTED SERPENT EAGLE *Spilornis cheela*
   For distribution see Plate 20

5. (202) GREAT NICOBAR SERPENT EAGLE *Spilornis klossi*
   For distribution see Plate 20

6. (202a) ANDAMAN DARK SERPENT EAGLE *Spilornis elgini*
   For distribution see Plate 20

7. (161) CRESTED HAWK-EAGLE *Spizaetus cirrhatus*
   For distribution see Plate 20

8. (125) BLYTH'S BAZA *Aviceda jerdoni*
   For distribution see Plate 22

9. (160) CHANGEABLE HAWK-EAGLE *Spizaetus cirrhatus limnaeetus*
   For distribution see Plate 20

10. (158) HODGSON'S HAWK-EAGLE *Spizaetus nipalensis*
    For distribution see Plate 20

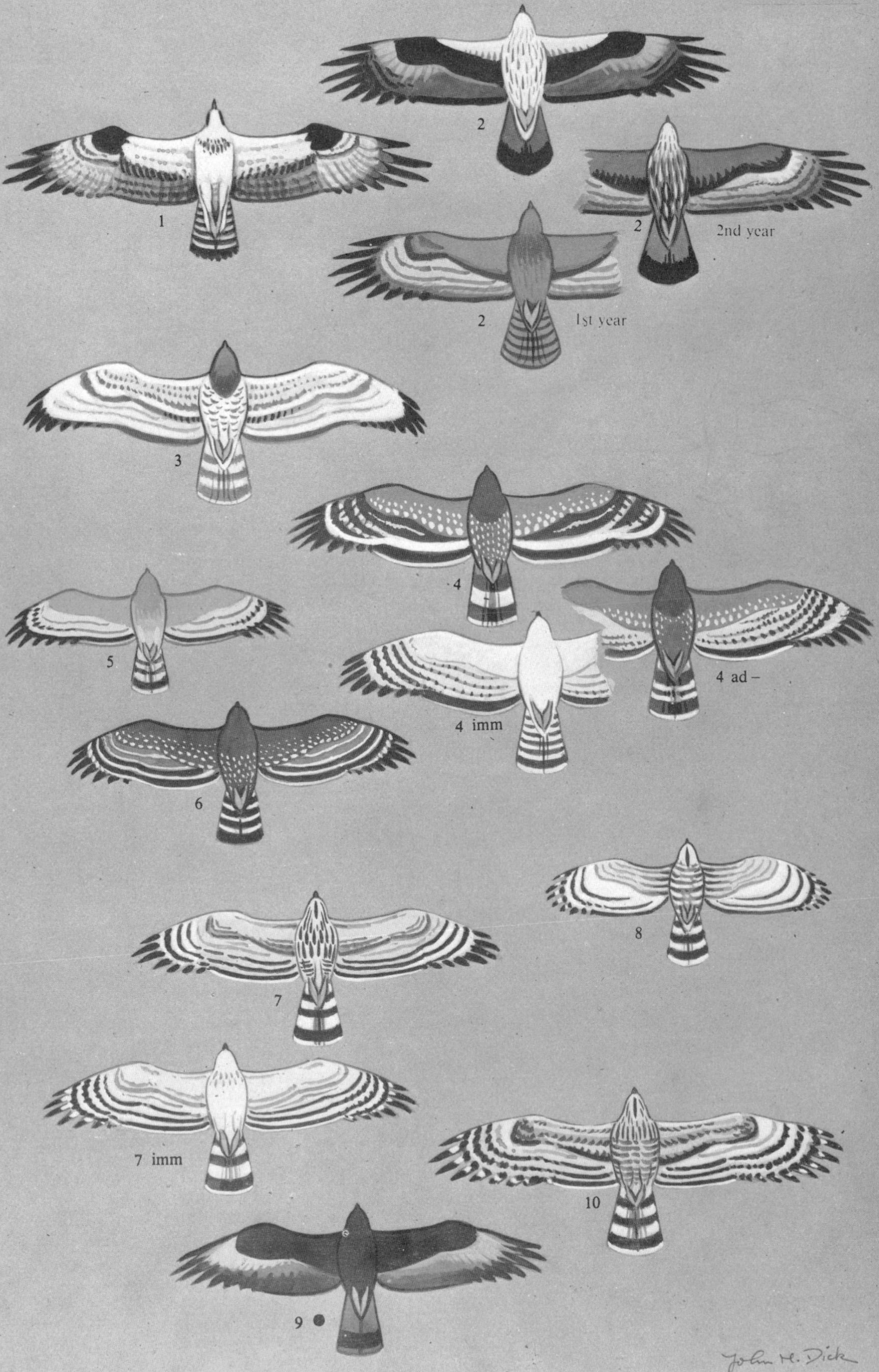

2

1

2

2 2nd year

2 1st year

3

4

5

4 ad –

4 imm

6

8

7

7 imm

10

9 ●

John H. Dick

PLATE 22
1 imm
1
2♂
3
3 imm
5♂
4 imm
4 ♂
5 ♀
5 imm
4 ♀
6 imm
6 ad –
6
6
7 imm
7 ♂
Accipiter flight style
9 ♂
9 ♀
8 ♀
8♂
8♂
11 imm
12 imm
11♂
12♂
10 imm
11♂
10
11♀
12♀
John H. Dick

PLATE 22

1. (124.) BLACKWINGED or BLACKSHOULDERED KITE *Elanus caeruleus* R Crow-: 33 cm. Deciduous biotope: thin savannah forest, grassland, cultivation; plains and hills up to *c.* 1300 m. Subcontinent; Sri Lanka.

2. (127) BLACKCRESTED BAZA *Aviceda leuphotes* ? RM Pigeon±: 33 cm. Tropical evergreen foothills forest. Range disjunct, more or less overlapping that of 125 q.v.

3. (125) BLYTH'S BAZA *Aviceda jerdoni* R Crow+: 48 cm. Tropical evergreen foothills. Range disjunct: E Himalayas (Sikkim, Darjeeling, Assam and associated states, Chittagong Hill Tracts) up to *c.* 1800 m; SW India (Kerala, Karnataka), Andhra Pradesh, Tamil Nadu and Sri Lanka, between *c.* 150 and 900 m.

4. (139) SHIKRA *Accipiter badius* R Crow-: 30-34 cm. Light deciduous forest, village groves, etc. Subcontinent; Bangladesh; Sri Lanka and Nicobars; plains and hills up to *c.* 1800.

5. (144) CRESTED GOSHAWK *Accipiter trivirgatus* R Crow+: 40-46 cm. Open deciduous and semi-evergreen foothills forest. Himalayas (up to *c.* 2000 m) from Garhwal east to Sikkim; Assam and associated states; south to the Godavari river; SW India from Goa, south to Kerala, Tamil Nadu; Sri Lanka.

6. (143) HORSFIELD'S GOSHAWK *Accipiter soloensis* M Crow-: 30 cm. Orissa (?) and Nicobar I.

7. (141) CAR NICOBAR SHIKRA *Accipiter badius butleri* R Crow-: 30 cm. Car Nicobar I.

8. (151) BESRA SPARROW-HAWK *Accipiter virgatus besra* R Crow-: 29-34 cm. Heavy evergreen and moist-deciduous forest. W. Ghats between *c.* 600 and 1200 m, from Bombay south through Kerala; sparingly also in E Ghats, Sri Lanka.

9. (152) EASTERN BESRA SPARROW-HAWK *Accipiter virgatus gularis* RM Crow-: 29-34 cm. Andaman Is. (breeding); also the Nicobars where breeding unconfirmed. Possible winter migrant to extreme E India and Bangladesh from E Asia. Collected once in Tamil Nadu in March.

10. (136) GOSHAWK *Accipiter gentilis* RM Kite±: M *c.* 50 cm, F *c.* 61 cm. Forests. Lower Himalayas from Kashmir to Arunachal Pradesh; breeding above *c.* 2400 m, in NW. Sporadic in Pakistan and peninsular India (winter).

11. (147) SPARROW-HAWK *Accipiter nisus nisosimilis* M. Crow-: 31-36 cm. Well-wooded country: light forest, groves, orchards, etc. Subcontinent. Absent in Sri Lanka.

12. (148) SPARROW-HAWK *Accipiter nisus melaschistos* RM Crow-: 31-36 cm. Forest and well-wooded country. Baluchistan and Himalayas, up to *c.* 4500 m. Descends to the foothills and adjacent plains in winter.

PLATE 23

1
2
3
3
4
4 imm
5 ♀
5 ♂
5 imm
6
6 imm
7
8
8
Accipiter flight style
9 ♀
9 ♂
10
11 ♂
11 ♀
11 imm
John H. Dick

PLATE 23

Raptors in flight (from below) contd.

1. (127) BLACKCRESTED BAZA *Aviceda leuphotes*
   For distribution see Plate 22

2. (125) BLYTH'S BAZA *Aviceda jerdoni*
   For distribution see Plate 21

3. (124) BLACKWINGED or BLACKSHOULDERED KITE *Elanus caeruleus*
   For distribution see Plate 22

4. (144) CRESTED GOSHAWK *Accipiter trivirgatus*
   For distribution see Plate 22

5. (139) SHIKRA *Accipiter badius*
   For distribution see Plate 22

6. (143) HORSFIELD'S GOSHAWK *Accipiter soloensis*
   For distribution see Plate 22

7. (152) EASTERN BESRA SPARROW-HAWK *Accipiter virgatus gularis*
   For distribution see Plate 22

8. (141) CAR NICOBAR SHIKRA *Accipiter badius butleri*
   For distribution see Plate 22

9. (151) BESRA SPARROW-HAWK *Accipiter virgatus*
   For distribution see Plate 22

10. (148) SPARROW-HAWK *Accipiter nisus*
    For distribution see Plate 22

11. (136) GOSHAWK *Accipiter gentilis*
    For distribution see Plate 22

PLATE 24

1. (130) HONEY BUZZARD *Pernis ptilorhyncus* RM Kite±: 68 cm. Semi-desert country, deciduous and semi-evergreen forest, and cultivation. Subcontinent; up to *c.* 1800 m in the Himalayas. Also Sri Lanka, Maldives. Six colour phases: a-f.

2. (156) JAPANESE BUZZARD *Buteo buteo japonicus* RM Kite-: 51-56 cm. Recorded in winter in E Himalayas, the hills south of the Brahmaputra to Manipur, and peninsular India south to Kerala and Tamil Nadu. Also Sri Lanka. *Breeding:* W Himalayas from Gilgit, Kashmir eastwards, above c. 2700 m.

3. (155) DESERT BUZZARD *Buteo buteo vulpinus* M Kite-: 51-56 cm. Easily confused with other buzzards due to the great variability in their plumages. Definite records from Nepal, Simla and Mishmi Hills. Five colour phases: a-e.

4. (EL) ROUGHLEGGED BUZZARD *Buteo lagopus* Holarctic.

5. (154) UPLAND BUZZARD *Buteo hemilasius* V Kite+: 71 cm. Rare winter visitor to the Himalayas. Specimens definitely identified only from Kashmir, Himachal Pradesh, Punjab, Nepal, Sikkim. Delhi (sight).

6. (153) LONGLEGGED BUZZARD *Buteo rufinus* RM Kite±: 61 cm. *Summer:* precipitous rocky ground with forest. *Winter:* open country and cultivation. Pakistan; Himalayas; southward in the Peninsula to Karnataka and Tamil Nadu; Sri Lanka, Three colour phases: a-c.

7. Typical buzzard flight style.

1a ●
1 c
1e
1 b
1d O
1 f
2 ●
2
2 O
3a
3 b
3 c
3 d
3e
4 imm
7 ↑
5
5 O
4
6 a
4 O
6 b ●
6 c O

John H. Dick

From below

PLATE 25

Kites, Buzzards and Eagles in flight (from below)

1. (130) HONEY BUZZARD *Pernis ptilorhyncus*
   For distribution see Plate 24

2. (135) BRAHMINY KITE *Haliastur indus*
   For distribution see Plate 14

3. (157) WHITE-EYED BUZZARD-EAGLE *Butastur teesa*
   For distribution see Plate 29

4. (155) DESERT BUZZARD *Buteo buteo vulpinus*
   For distribution see Plate 24

5. (156) JAPANESE BUZZARD *Buteo buteo japonicus*
   For distribution see Plate 24

6. Typical buzzard flight style

7. (EL) ROUGHLEGGED BUZZARD *Buteo lagopus*
   For distribution see Plate 24

8. (153) LONGLEGGED BUZZARD *Buteo rufinus*
   For distribution see Plate 24

9. (154) UPLAND BUZZARD *Buteo hemilasius*
   For distribution see Plate 24

10. (165) RUFOUSBELLIED HAWK-EAGLE *Hieraaetus kienerii*
    For distribution see Plate 20

11. (164) BOOTED HAWK-EAGLE *Hieraaetus pennatus*
    For distribution see Plate 17

PLATE 26

1. (172) BLACK EAGLE *Ictinaetus malayensis* R Kite+: 69-81 cm. Partial to hill forest. Lower Himalayas (up to *c.* 2700 m); hills south of the Brahmaputra; continental and peninsular India south to Kanyakumari Dt.; Sri Lanka.

2. (171) LESSER SPOTTED EAGLE *Aquila pomarina* R Kite+: 61-66 cm. Open wooded country and cultivation. Chiefly the Gangetic plain and Nepal terai, east through Bangladesh to Burma, south to Gujarat, Madhya Pradesh, Orissa and Bombay.

3. (170) GREATER SPOTTED EAGLE *Aquila clanga* RM Kite+: 64-72 cm. Large marshes, jheels and canals. Pakistan, N India and Nepal, east through Bangladesh to Burma, south to *c.* 20°N latitude. Variant plumages: a-d.

4. (168) TAWNY EAGLE *Aquila rapax vindhiana* R Kite+: 63-71 cm. Semi-desert and dry- and moist-deciduous country. Baluchistan and NWFP east to Nepal terai and Bangladesh, south to Kanara and Tamil Nadu. Variant immatures: a-c.

5. (169) EASTERN STEPPE EAGLE *Aquila rapax nipalensis* M Kite+: 76-80 cm. Open plains, often near water. Pakistan and N India, Bangladesh (?), south to Bombay and Orissa.

6. (166) GOLDEN EAGLE *Aquila chrysaetos* R Vulture±: 90-100 cm. Rugged high mountain country between 1800 m and 5000+ m. Baluchistan and Himalayas.

7. (167) IMPERIAL EAGLE *Aquila heliaca* RM Vulture-: 81-90 cm. Open treeless country. Pakistan and NW India. In winter south to Saurashtra (Gujarat), east to Nepal, Bangladesh. Maharashtra (once).

Variant plumages
1 imm
1
2
2 imm
3 d
3 c
3 b
3
3a
5 ad –
4c
4b
4a
5 ad –
5
4 ad –
4
4 ad –
6 ad –
6
7
7 ad –
7 ad –
John M. Dick

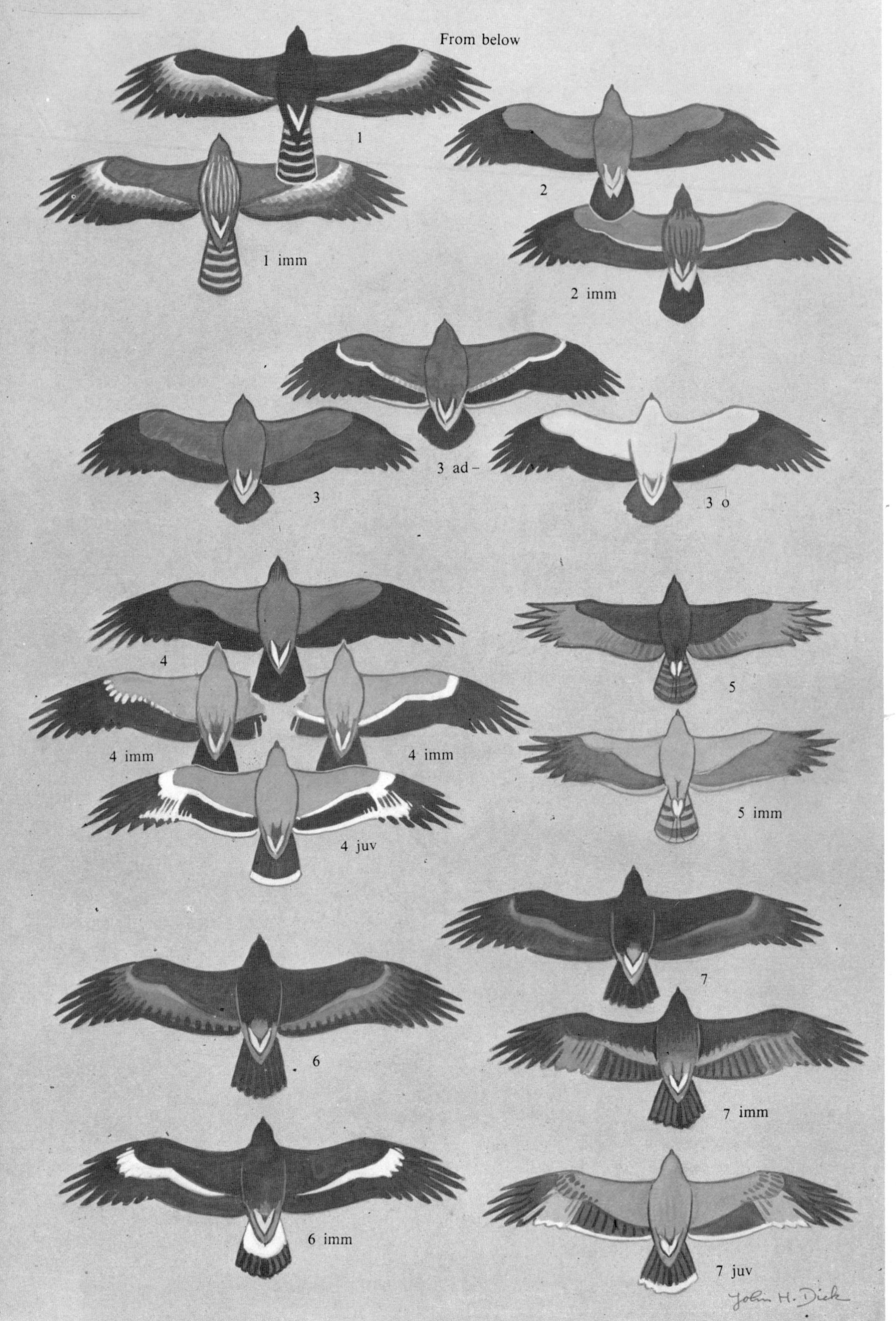
From below
1
1 imm
2
2 imm
3 ad –
3
3 o
4
4 imm
4 imm
4 juv
5
5 imm
7
6
7 imm
6 imm
7 juv
John H. Dick

PLATE 27

Eagles in flight (from below)

1. (172) BLACK EAGLE *Ictinaetus malayensis*
   For distribution see Plate 26

2. (171) LESSER SPOTTED EAGLE *Aquila pomarina*
   For distribution see Plate 26

3. (170) GREATER SPOTTED EAGLE *Aquila clanga*
   For distribution see Plate 26

4. (169) EASTERN STEPPE EAGLE *Aquila rapax nipalensis*
   For distribution see Plate 26

5. (168) TAWNY EAGLE *Aquila rapax vindhiana*
   For distribution see Plate 26

6. (166) GOLDEN EAGLE *Aquila chrysaetos*
   For distribution see Plate 26

7. (167) IMPERIAL EAGLE *Aquila heliaca*
   For distribution see Plate 26

PLATE 28

Baza, Kites, Buzzards & Eagles in flight

1. (127) INDIAN BLACKCRESTED BAZA *Aviceda leuphotes*
   For distribution see Plate 22

2. (124) BLACKWINGED or BLACKSHOULDERED KITE *Elanus caeruleus*
   For distribution see Plate 22

3. (157) WHITE-EYED BUZZARD-EAGLE *Butastur teesa*
   For distribution see Plate 29

4. (196) CRESTED SERPENT EAGLE *Spilornis cheela*
   For distribution see Plate 20

5. (EL) ROUGHLEGGED BUZZARD *Buteo lagopus*
   For distribution see Plate 24

6. (169) EASTERN STEPPE EAGLE *Aquila rapax nipalensis*
   For distribution see Plate 26

7. (170) GREATER SPOTTED EAGLE *Aquila clanga*
   For distribution see Plate 26

8. (168) TAWNY EAGLE *Aquila rapax vindhiana*
   For distribution see Plate 26

9. (166) GOLDEN EAGLE *Aquila chrysaetos*
   For distribution see Plate 26

10. (167) IMPERIAL EAGLE *Aquila heliaca*
    For distribution see Plate 26

11. (163) BONELLI'S EAGLE *Hieraaetus fasciatus*
    For distribution see Plate 20

12. (161) CRESTED HAWK-EAGLE *Spizaetus cirrhatus*
    For distribution see Plate 20

1♂
1♀
2
3
4
5
6
6 imm
7
7 imm
7 imm
8
9
10 imm
9 imm
11
12

PLATE 29
1
2 imm
2
3 imm
3
4 ♀
4 ♂
5 imm
5 ♀
5 ♂
6 ♀
6 ♂
8
7 imm
9 ♀
9 ♂
7
10 imm
10
11 imm
11 ad –
11 ○
11 ●
12 imm
12
13 imm
13
14 imm
14 ad –
14
15 imm
15 ○
15 ●
16 imm
16
17
John H. Dick

PLATE 29

1. (205) WHITELEGGED FALCONET *Microhierax melanoleucos* R Bulbul±: 20 cm. Outskirts of forest and cultivation clearings up to *c.* 1500 m. Assam and the hill states south of the Brahmaputra, Bangladesh.

2. (204) REDBREASTED FALCONET *Microhierax caerulescens* R Bulbul±: 18 cm. Outskirts of forest and cultivation clearings up to *c.* 2000 m. Himalayan foothills from Kumaon east, Assam, Orissa (sight).

3. (157) WHITE-EYED BUZZARD-EAGLE *Butastur teesa* R Crow±: 43 cm. Open dry forest, scrub and cultivation, Subcontinent. Absent Sri Lanka.

4. (221) LESSER KESTREL *Falco naumanni* M Pigeon±: 34 cm. Open savannah country. From Gilgit, Haryana and Nepal to Assam and associated states, south to Maharashtra, Tamil Nadu and Orissa. Perhaps an irregular passage migrant.

5. (220) REDLEGGED FALCON *Falco vespertinus* (Passage) M Pigeon-: 28-31 cm. Open country and grazing land. NE India, Bangladesh, Sikkim, Bhutan (?), Nepal, Gujarat, the Indian peninsula and Maldive I. (winter). Sri lanka (straggler). Breeds (casually) Assam.

6. (222) KESTREL *Falco tinnunculus* RM Pigeon±: 36 cm. Grassland, semi-desert, cultivation and rocky regions. *Breeding:* From NWFP to Baluchistan and Punjab (Pakistan); Himalayas; W and E Ghats; Sri Lanka. *Winter:* Subcontinent.

7. (212) HOBBY *Falco subbuteo* RM Pigeon±: 31-34 cm. Open wooded country and semi-desert. Pakistan, Nepal, Bangladesh and India south to Belgaum.

8. (219) REDHEADED MERLIN *Falco chicquera* R Pigeon±: 31-36 cm. Open country with groves of trees and cultivation. Subcontinent.

9. (217) MERLIN *Falco columbarius* M Pigeon-: 27-30cm. Scrubby, open country and cultivation. Pakistan and Nepal (winter). Ladakh (passage).

10. (215) ORIENTAL HOBBY *Falco severus* R Pigeon-: 27-30 cm. Well-wooded foothills country. Himalayas; Bangladesh; NE India; *c.* 1800-2400 m. Calcutta and Kerala (winter); Sri Lanka (straggler).

11. (216) SOOTY FALCON *Falco concolor* V Crow-: 38 cm. Single record from the coast of Baluchistan.

12. (209) PEREGRINE FALCON *Falco peregrinus japonensis* M Kite-; Crow+: 40-48 cm. Near rivers, marshes and lagoons. Subcontinent, Lakshadweep, Andamans (?), Nicobars (?); Sri Lanka (winter).

13. (211) SHAHEEN FALCON *Falco peregrinus peregrinator* R Kite-; Crow+: 38-46 cm. Rugged hilly country. Subcontinent. Also Sri Lanka and Nicobars.

14. (208) LAGGAR FALCON *Falco biarmicus jugger* RM Kite-: 43-46 cm. Dry open country and cultivation. Subcontinent. Absent Sri Lanka.

15. (EL) BARBARY FALCON *Falco pelegrinoides* N Africa.

16. (206) SAKER or CHERRUG (LANNER) FALCON *Falco biarmicus cherrug* M Kite-: 50-56 cm. Desert and semi-desert. Pakistan and NW India in Ladakh, Haryana and Rajasthan; also recorded in Delhi district and Madhya Pradesh (winter).

17. (207) SHANGHAR FALCON *Falco biarmicus milvipes* M Kite-: 50-58 cm. Pakistan, NW India, Nepal (winter). Rare.

PLATE 30

Falcons in flight (from below)

1. (205) WHITELEGGED FALCONET *Microhierax melanoleucos*
For distribution see Plate 29

2. (221) LESSER KESTREL *Falco naumanni*
For distribution see Plate 29

3. (204) REDBREASTED FALCONET *Microhierax caerulescens*
For distribution see Plate 29

4. (222) KESTREL *Falco tinnunculus*
For distribution see Plate 29

5. (220) REDLEGGED FALCON *Falco vespertinus*
For distribution see Plate 29

6. (217) MERLIN *Falco columbarius*
For distribution see Plate 29

7. (219) REDHEADED MERLIN *Falco chicquera*
For distribution see Plate 29

8. (216) SOOTY FALCON *Falco concolor*
For distribution see Plate 29

9. (215) ORIENTAL HOBBY *Falco severus*
For distribution see Plate 29

10. (212) HOBBY *Falco subbuteo*
For distribution see Plate 29

11. (206) LANNER (SAKER or CHERRUG) FALCON *Falco biarmicus cherrug*
For distribution see Plate 29

12. (211) SHAHEEN FALCON *Falco peregrinus peregrinator*
For distribution see Plate 29

13. (209) PEREGRINE FALCON *Falco peregrinus japonensis*
For distribution see Plate 29

14. (208) LAGGAR FALCON *Falco biarmicus jugger*
For distribution see Plate 29

PLATE 30

1
2 ♂
3 imm
4 ♂
2 ♀
3
4
4 ♀
5 ♂
7
5 ♀
6 ♂
6 ♀
8
9
10
8 imm
9 imm
11
12
11 imm
12 imm
14
14 ad
13
14 imm
13 imm

PLATE 31
1♂
♀
2
2
3♂
♀
4 ♂
♀
5
6 ♂
♀
10
7
8
9
13 ♂
♀
11 ♂
♀
12 ♂
♀
John H. Dick

PLATE 31

1. (238) BLACK PARTRIDGE *Francolinus francolinus* R 34 cm. Well-watered tracts with tall grass and scrub, sugarcane fields and wet cultivation. Pakistan and N India from Punjab and Kashmir east through Nepal to Bangladesh, Bhutan, Assam and associated NE States; from Himalayan foothills south to N Gujarat, N Madhya Pradesh and N Orissa.

2. (246) GREY PARTRIDGE *Francolinus pondicerianus* R 33 cm. Thorn-Scrub, dry light jungle, cultivation environs. Subcontinent; Sri Lanka. Introduced to Andaman I.

3. (243) CHINESE FRANCOLIN *Francolinus pintadeanus* R 33 cm. Low scrub and grassy openings in forest. SE Manipur State.

4. (241) PAINTED PARTRIDGE *Francolinus pictus* R 31 cm. Dry grass and scrub. N, C and peninsular India, Sri Lanka (now very rare). Absent in humid tracts of Karnataka and Kerala.

5. (247) SWAMP PARTRIDGE or KYAH *Francolinus gularis* R Partridge+: 37 cm. Swampy grass jungle. N India from the Kumaon terai and Uttar Pradesh, east through N Bihar, Bangladesh, Sunderbans and NE India.

6. (267) COMMON HILL PARTRIDGE *Arborophila torqueola* R Partridge-: 28 cm. Dense evergreen jungle. Himalayas from Chamba east; Assam and associated States; between 600 and 4000 m.

7. (272) WHITENECKED HILL PARTRIDGE *Arborophila atrogularis* R Partridge-: 28 cm. Undergrowth in wet evergreen forest and bamboo jungle. From Arunachal Pradesh south through the hills south of the Brahmaputra to Sylhet and the Chittagong region (Bangladesh), up to *c.* 1500 m.

8. (273) REDBREASTED HILL PARTRIDGE *Arborophila mandellii* R Partridge-: 28 cm. Undergrowth in evergreen forest. Sikkim, Bhutan and Arunachal Pradesh; 350-2500 m.

9. (270) RUFOUSTHROATED HILL PARTRIDGE *Arborophila rufogularis* R Partridge-: 27 cm. Dense secondary scrub and undergrowth in evergreen forest. Himalayas from Kumaon east, seasonally between *c.* 700 and 2400 m.

10. (271) SSP *intermedia* of 270. NE hill states; Bangladesh Chittagong). *c.* 600-1800 m.

11. (275) RED SPURFOWL *Galloperdix spadicea* R Partridge+: 36 cm. Dry- and moist-deciduous scrub, in broken hilly country; up to *c.* 1250 m. From Rajasthan, Uttar Pradesh, Nepal terai and Bihar, south to Andhra Pradesh (E. Ghats), Tamil Nadu, Karnataka and Kerala; Gujarat.

12. (279) CEYLON SPURFOWL *Galloperdix bicalcarata* R Partridge+: 34 cm. Dense forest. Wet and dry zones, up to *c.* 2000 m. Sri Lanka.

13. (278) PAINTED SPURFOWL *Galloperdix lunulata* R 32 cm. Dense thorn scrub and bamboo jungle in broken terrain and rocky foothills. India south of the Gangetic plain from *c.* Gwalior in the west and Bengal in the east; up to *c.* 1000 m.

PLATE 32
1 ♀
♂
2 ♂
3 ♀
♀
4 ♂
♂
5 ♀
6 ♂
♀
♀
7 ♂
♀
8 ♂
♀
9 ♂
♀
10 ♂
♀
11 ♂
12 ♂
♀
13 ♂
14
15
16
17
18
John H. Dick

PLATE 32

1. (313) LITTLE BUSTARD-QUAIL *Turnix sylvatica* R Quail-: Male 13 cm; Female slightly larger. Grass-and-scrub jungle. From the Punjab Salt Range to Nagaland, Manipur and Chittagong, and from Nepal and Sikkim south to Kerala. Sri Lanka? Up to *c.* 2400 m.

2. (318) COMMON BUSTARD-QUAIL *Turnix suscitator* R Quail-; Male 15 cm. Female slightly larger. Grassland and scrub jungle and open light deciduous forest. Entire Subcontinent in suitable biotope.

3. (319) Sri Lanka ssp. *leggei* of 318.

4. (253) BLUEBREASTED QUAIL *Coturnix chinensis* R Quail-: 14 cm. Swampy grassland and grass-and-scrub. Subcontinent east and south of a line Bombay to Simla; Sri Lanka; Nicobar I.

5. (314) YELLOWLEGGED BUTTON QUAIL *Turnix tanki* R Quail-: Male 15 cm; Female slightly larger. Grassland with scrub, bamboo jungle and standing crops. Entire Subcontinent east of Sind and NWFP of Pakistan. Also Andaman and Nicobar Is.

6. (250) GREY QUAIL *Coturnix coturnix* RM Partridge-: 20 cm. Grassland and standing crops. *Breeding:* Pakistan east through N India and Bangladesh, from Himalayan foothills south to Maharashtra. *Winter*: Rest of Subcontinent; not Sri Lanka. Race *japonica* winters in NE India.

7. (263) PAINTED BUSH QUAIL *Perdicula erythrorhyncha* R Quail-: 16 cm. Low hilly deciduous biotope. Bihar, W Bengal, Orissa, W Madhya Pradesh and E Maharashtra; W Ghats from Khandala southward to Kerala, E Ghats (Andhra) and Tamil Nadu.

8. (252) BLACKBREASTED or RAIN QUAIL *Coturnix coromandelica* R Quail-: 18 cm. Grass and scrub, moist grassland and standing crops. Subcontinent up to *c.* 2000 m. Sri Lanka occasional, winter.

9. (265) MANIPUR BUSH QUAIL *Perdicula manipurensis* R Quail±: 20 cm. Damp grasslands. N Bengal, NE Bangladesh, Assam and the hills south of the Brahmaputra, from Meghalaya and Nagaland south to Manipur and the mountain ranges of Chittagong.

10. (261) ROCK BUSH QUAIL *Perdicula argoondah* R Quail-: 17 cm. Open semi-desert and thorn-scrub country, preferably drier and stonier facies than Jungle Bush Quail's but often overlapping. Practically entire Subcontinent; not Sri Lanka.

11. (255) JUNGLE BUSH QUAIL *Perdicula asiatica* R Quail-: 17 cm. Grass-and-scrub jungle and open deciduous secondary forest. Practically entire Subcontinent, plains and hills up to *c.* 1250 m. Sri Lanka.

12. (280) MOUNTAIN QUAIL *Ophrysia superciliosa* Extinct? Partridge-: 25 cm. Grass and brushwood on steep Himalayan hillsides in the Dehra Dun-Naini Tal region, *c.* 1650-2000 m. Rare; last reported in 1876.

13. (228) SEESEE PARTRIDGE *Ammoperdix griseogularis* R Partridge-: 26 cm. Stony foothils. Pakistan from Baluchistan and NWFP to the Salt Range, and the Kirthar hills of Sind.

14. (227) SNOW PARTRIDGE *Lerwa lerwa* R Partridge+: 38 cm. Alpine meadows, open bush-covered hillsides above tree-line. Himalayas, *c.* 2000 and 5000 m.

15. (274) BAMBOO PARTRIDGE *Bambusicola fytchii* R Partridge+: 35 cm. Open scrub and edges of cultivation. The hills south of the Brahmaputra from the Lakhimpur district south to Sylhet and the Chittagong region. Up to *c.* 2000 m.

16. (236) CHUKAR PARTRIDGE *Alectoris chukar* R Partridge+: 38 cm. Rocky hillsides with grass and sparse bushes. Pakistan mountains and W Himalayas, east to C Nepal, 1000-5000 m.

17. (249) TIBETAN PARTRIDGE *Perdix hodgsoniae* R Partridge±: 31 cm. Furze, dwarf juniper and rhododendron scrub on stony hillsides. Ladakh, Kumaon, Nepal east to Bhutan to the Tsangpo bend, *c.* 3500-5500 m.

18. (225) MEGAPODE *Megapodius freycinet* R Village hen±: 43 cm. Undergrowth in dense forest fringe and sandy seashore above spring-tide mark, Andaman (old records only) and Nicobar I.

PLATE 33

1. (311) COMMON PEAFOWL *Pavo cristatus* R Village hen+: Male *c.* 92-122 cm without train, *c.* 2-2.5 m in full plumage; Female 86 cm. Moist-and dry-deciduous forest, cultivation and precincts of villages. Subcontinent east of the Indus, up to *c.* 1800 m; Sri Lanka. Introduced in Sind and Andaman I.

2. (312) BURMESE PEAFOWL *Pavo muticus* R Village hen+: Male *c.* 92-122 cm without train, *c.* 2-2.5 m in full plumage; Female 86 cm. Dense evergreen and moist-deciduous forest. NE hill states, Bangladesh (Chittagong region) and N Bengal; up to *c.* 1000 m.

1 ♂
1 ♂
1 ♀
1 ♂ moulting
1 ♂ in display
2 ♀
2 ♂
John H. Dick

PLATE 34

PLATE 34

1. (285) WESTERN TRAGOPAN *Tragopan melanocephalus* R Village hen±: 71 cm. Dense rhododendron, ringal bamboo and other undergrowth in conifer or oak forest. W Himalayas from Swat in Pakistan east through Kashmir, Ladakh, Himachal Pradesh and Garhwal; seasonally between *c.* 1350 and 3600 m.

2. (286) CRIMSON or SATYR TRAGOPAN *Tragopan satyra* R Village hen±: M 68 cm; Female 59 cm. Rhododendron, bamboo and other undergrowth in conifer or oak forest. Himalayas from Garhwal east to Bhutan and W Arunachal Pradesh; seasonally between *c.* 2000 and 4500 m.

3. (233) PHEASANT-GROUSE *Tetraophasis szechenyii* R Village hen±: 64 cm. Rocky scrub-covered ravines. Subalpine zone. N Arunachal Pradesh *c.* 3400-4600 m

4. (289) TEMMINCK'S TRAGOPAN *Tragopan temminckii* R Village hen±: 64 cm. Dense undergrowth in evergreen hill forest. Arunachal Pradesh, *c.* 2100-3500 m.

5. (288) BLYTH'S or GREYBELLIED TRAGOPAN *Tragopan blythii* R Village hen±: Male 68 cm; Female 59 cm. Dense undergrowth in evergreen forest. E Bhutan, east to Arunachal Pradesh hills south of the Brahmaputra from the Patkai range through Nagaland and Manipur to Mizoram; seasonally between *c.* 1800 and 3300 m.

6. (290) IMPEYAN or MONAL PHEASANT *Lophophorus impejanus* R Village hen+: 72 cm. Rhododendron and other undergrowth in open conifer forest interspersed with grassy glades, and the scrub zone above timber-line. Himalayas, seasonally *c.* 1800-5000 m.

7. (291) SCLATER'S MONAL *Lophophorus sclateri* R Village hen+: 72 cm. Dense rhododendron undergrowth in conifer forest. Subalpine zone of Arunachal Pradesh between 92° and 93°E long.; *c.* 3000-4000 m.

8. (229) TIBETAN SNOWCOCK *Tetraogallus tibetanus* R Village hen+: 70 cm. Stony slopes and alpine meadows above timber-line. Himalayas from Ladakh and Rupshu east, up to 5800 m.

9. (232) HIMALAYAN SNOWCOCK *Tetraogallus himalayensis* R Village hen+: 72 cm. Stony slopes and alpine meadows above timber-line. W Himalayas, seasonally between *c.* 2400 and 5500 m.

10. (292) EARED PHEASANT *Crossoptilon crossoptilon* R Village hen+: 72 cm. Rhododendron and juniper undergrowth in conifer forest interspersed with grassy hill-slopes. Arunachal Pradesh Himalayas; *c.* 3000-5000 m.

PLATE 35

1. (304) KOKLAS PHEASANT *Pucrasia macrolopha* R Village hen±: M 61 cm; F 53 cm. Wooded slopes and nullahs in oak and conifer forest. Himalayas from NWFP (Pakistan) eastward to W Nepal; seasonally between *c.* 1000 and 4200 m.

2. (307) CHIR PHEASANT *Catreus wallichii* R Village hen+: M 90-118 cm. F 61-76 cm. Open conifer and deciduous forest interspersed with grassy slopes. W Himalayas, east to W-C Nepal, *c.* 1500-2700 m.

3. (EL) COMMON PHEASANT *Phasianus colchicus* Temperate Asia. Introduction being tried in Bhutan.

4. (294) KALIJ PHEASANT *Lophura leucomelana* R Village hen±: 60-68 cm. Foothills forest with heavy scrub undergrowth; partial to neighbourhood of water and terraced cultivation. Himalayas; NE hill states; Bangladesh: *c.* 300-3600 m.

5. (282) BLOOD PHEASANT *Ithaginis cruentus* R Village hen±: 46 cm. Dwarf rhododendron, juniper and ringal bamboo undergrowth in conifer forest. Subalpine zone of E Himalayas, seasonally between *c.* 1500 and 4700 m.

6. (299) RED JUNGLEFOWL *Gallus gallus* R Village hen±: Male 66 cm; Female 43 cm. Moist-deciduous forest and scrub jungle interspersed with cultivation. Himalayan foothills; NE hill states; N and NC India; and E Ghats (Wangasara); Bangladesh.

7. (308) MRS HUME'S BARREDBACK PHEASANT *Syrmaticus humiae* R Village hen+: M 90 cm; F 60 cm. Open forest interspersed with grassy and bracken covered slopes. NE hill states, from the Patkai Hills and Barail Range south through Nagaland and Manipur to Mizoram; *c.* 900-2000 m.

8. (302) CEYLON JUNGLEFOWL *Gallus lafayettii* Endemic. Village hen±: M 66-72 cm; F 36 cm. All types of forest; lowland to highest hills. Sri Lanka.

9. (301) GREY or SONNERAT'S JUNGLEFOWL *Gallus sonneratii* R Village hen±: M 60-80 cm; F 46 cm. Dry-deciduous to moist-evergreen biotope: broken country interspersed with cultivation patches. Chiefly W peninsular India and south of a line Mt Abu-Pachmarhi-mouth of the Godavari.

10. (310) PEACOCK-PHEASANT *Polyplectron bicalcaratum* R Village hen±: M 64 cm; F 48 cm. Dense evergreen jungle, duars and foothills. E Himalayas from Sikkim through Arunachal Pradesh; NE hill states, south to the Chittagong hill tracts (Bangladesh).

♀
1♂
♀
2♂
♀
3♂
♀
4♂
♀
5♂
♀
6♂
7
♀
♀
8♂
♀
9♂
♀
10♂
John M. Dick

PLATE 36

1
1
2
3
4 ♂
4 ♀
5
6
7
8
9
10
11
12
13 ♂
♀
14
15
15
16
16 br –
16 br +
17
18
18 imm
19
John H. Dick

PLATE 36

1. (327) WATER RAIL *Rallus aquaticus* RM Partridge-: 28 cm. Reedy marshes. *Breeding:* Kashmir and Ladakh(?). *Winter:* Pakistan, NW and N India to Assam; NE India; Bangladesh, Madhya Pradesh.

2. (329) BLUEBREASTED BANDED RAIL *Rallus striatus* R Partridge-: 27 cm. Reedy swamps, mangroves, paddyfields. Subcontinent minus Pakistan; Andaman and Nicobar Is; Sri Lanka.

3. (331) REDLEGGED BANDED CRAKE *Rallina fasciata* R (?) Quail+: 23 cm. Dense forest. Recorded only from Assam (N Cachar).

4. (335) LITTLE CRAKE *Porzana parva* M Quail±: 20 cm. Swampy reed-beds. Pakistan and W Peninsular India.

5. (332) BANDED CRAKE *Rallina eurizonoides* RM Quail+: 25 cm. Dense jungle in well-watered country. Subcontinent. Sri Lanka (winter).

6. (337) BAILLON'S CRAKE *Porzana pusilla* RM Quail±: 19 cm. Swampy reed-beds, irrigated crops, etc. *Breeding:* Kashmir, Uttar Pradesh, Nepal. *Winter:* Subcontinent, Andamans; Sri Lanka.

7. (334) CORN CRAKE *Crex crex* V Partridge-: 25 cm. Only four records. Gilgit, Sri Lanka.

8. (338) SPOTTED CRAKE *Porzana porzana* M Quail+: 23 cm. Marshy reed-beds. Subcontinent except extreme south.

9. (339) RUDDY CRAKE *Porzana fusca* R Quail+: 22 cm. Marshes, flooded ricefields, etc. Subcontinent; Sri Lanka.

10. (342) BROWN CRAKE *Amaurornis akool* R Partridge-: 28 cm. Marshes and grassy margins of streams. From Kashmir east to Assam and associated states; Manipur; Bangladesh; peninsular India minus extreme south.

11. (341) ELWES'S CRAKE *Amaurornis bicolor* R Quail+: 22 cm. Swampy dense grass jungle around ponds and streams. Nepal to Arunachal Pradesh; NE hill states south to Bangladesh; from foothills to *c.* 2800 m.

12. (333) ANDAMAN BANDED CRAKE *Rallina canningi* R Partridge+: 34 cm. Marshy jungle. Andaman Is.

13. (346) WATER COCK *Gallicrex cinerea* R Partridge+: M 43 cm; F 36 cm. Marshes, ponds, ricefields, etc. Subcontinent east of the Indus, Andamans and Nicobars, Sri Lanka, Maldives.

14. (347) MOORHEN *Gallinula chloropus* RM Partridge+: 32 cm. Jheels, marshes, ponds, etc. Subcontinent, Andamans; Sri Lanka.

15. (350) COOT *Fulica atra* RM Three-quarter grown duck±: 42 cm. Large jheels, lakes and reservoirs. Subcontinent.

16. (358) PHEASANT-TAILED JACANA *Hydrophasianus chirurgus* R Partridge±: 31 cm (excluding tail). Jheels and ponds with floating vegetation. Subcontinent; Sri Lanka.

17. (343) WHITEBREASTED WATERHEN *Amaurornis phoenicurus* R Partridge±: 32 cm. Marshes, ponds and ricefields. Subcontinent, Andamans and Nicobars; Sri Lanka; Maldives.

18. (359) BRONZEWINGED JACANA *Metopidius indicus* R Partridge±: Male 28 cm; Female 31 cm. Jheels and ponds with floating vegetation. Subcontinent.

19 (349) PURPLE MOORHEN *Porphyrio porphyrio* R Village hen±: 43 cm. Marshy reed-beds, flooded ricefields, etc. Subcontinent. Sri Lanka.

PLATE 37

1. (353) LITTLE BUSTARD *Otis tetrax* M Village hen±: 46 cm. Grassland and cultivation. Mainly NW Pakistan. Straggler for rest of Pakistan and NW India.

2. (352) GREAT BUSTARD *Otis tarda* V Vulture+: 102 cm. Cultivation. Pakistan (Chitral, Peshawar, N Baluchistan and N Sind). Less than a dozen records.

3. (356) BENGAL FLORICAN *Eupodotis bengalensis* R Village hen±: standing 55 cm; Male 66 cm; Female slightly larger. Low lying wet grassland with bushes. Kumaon and Nepal terai, N Bengal and Bhutan duars, Arunachal Pradesh, Assam, Meghalaya, Bangladesh (Chittagong region).

4. (355) HOUBARA *Chlamydotis undulata* RM Village hen±: standing *c.* 60 cm; Male 73 cm; Female 66 cm. Sandy semi-desert. *Breeding:* Baluchistan. *Winter:* the rest of Pakistan, Rajasthan, Kutch and Saurashtra. Straggler to Kashmir, Delhi and Kerala.

5. (354) GREAT INDIAN BUSTARD *Choriotis nigriceps* R Vulture+: Male 122 cm; Female 92 cm. Grassland with scrub, and semi-desert. Formerly from the Indus river to Bengal and south to Tamil Nadu; now chiefly Rajasthan, Gujarat, Maharashtra, Madhya Pradesh and the Deccan south to Karnataka, Andhra Pradesh.

6. (357) LIKH or LESSER FLORICAN *Sypheotides indica* R Village hen±: Male 46 cm; Female 51 cm. Grassland with bushes and cultivation. Pakistan and India, seasonally in Sind, Punjab, Rajasthan, Gujarat, Madhya Pradesh, south to Karnataka, Kerala and Tamil Nadu.

1 imm
1♂ br +
2 ♂
1♀
1♂ br +
3 ♂
2 ♀
3 ♀
2 ♂
2♀
3 ♀
3 ♂ br +
4
5♂
5♀
4
6♂
6 imm
6♀
♀
5♀
6 ♂ br +
5 ♂
John H. Dick

PLATE 38
1
2
3
4
5
6 W
7 W
8 ♂
6 S
7 S
8 ♀
9 W
10
9 S
11
12 imm
12
13
14
15
15
15 imm

PLATE 38

1. (439) CREAMCOLOURED COURSER *Cursorius cursor* M Partridge-: 23 cm. Desert and semi-desert. Pakistan, Rajasthan and W Gujarat.

2. (440) INDIAN COURSER *Cursorius coromandelicus* R Partridge-: 26 cm. Dry stony plains and fallow land. Subcontinent; Sri Lanka.

3. (441) JERDON'S or DOUBLEBANDED COURSER *Cursorius bitorquatus* Extinct (?). Partridge-: 27 cm. Rocky ground with scrub jungle. Formerly the Penner and Godavari valleys in Andhra Pradesh. Last record in 1900. Rediscovered in Lankamalai Sanctuary, Cuddapah Dt. (AP) on 12 January 1986.

4. (443) COLLARED PRATINCOLE *Glareola pratincola* RM Myna±: 24 cm. Flood-plains of rivers, and near jheels. *Breeding:* Baluchistan, Sind, N India to W Bengal , Assam; Bangladesh; Sri Lanka. *Winter:* Subcontinent.

5. (444) SMALL INDIAN PRATINCOLE *Glareola lactea* R Sparrow+: 17 cm. Large streams, marshes and coastal swamps, Subcontinent east of the Indus river.

6. (371) GREY or BLACKBELLIED PLOVER *Pluvialis squatarola* M Partridge±: 31 cm. Sandy seashores, tidal creeks and mudflats; less commonly inland. Subcontinent.

7. (373) EASTERN GOLDEN PLOVER *Pluvialis dominica* M Quail±: 19 cm. Muddy shores, ploughed fields and wet grounds. Subcontinent; also Lakshadweep, Andamans and Nicobars; Sri Lanka.

8. (429) PAINTED SNIPE *Rostratula benghalensis* R Quail+: 25 cm. Reedy marsshes and pools. Subcontinent; Sri Lanka.

9. (364) LAPWING *Vanellus vanellus* M Partridge±: 31 cm. Fallow land, stubble and marshland bordering cultivation. Pakistan and NW India east to U.P., Nepal, Bangladesh, Assam and Manipur.

10. (369) SPURWINGED LAPWING *Vanellus spinosus* R Partridge±: 31 cm. Sandbanks and shingle beds of rivers. Manipur, Assam and associated states, Bangladesh, Bhutan, Nepal and N India west to Haryana; Madhya Pradesh south to the Godavari river.

11. (366) REDWATTLED LAPWING *Vanellus indicus* R Partridge+: 33 cm. Near water in open country and cultivation. Subcontinent; Sri Lanka.

12. (365) GREYHEADED LAPWING *Vanellus cinereus* M Partridge+: 37 cm. Wet ground, marshes, ploughed fields and stubble. Nepal valley, Bihar, Bangladesh, Assam and Manipur. Straggler to Kashmir, Rajasthan, Goa, Karnataka and Andamans.

13. (370) YELLOW-WATTLED LAPWING *Vanellus malabaricus* R Partridge-: 27 cm. Waste land, stubbles and fallow fields in dry biotope. Subcontinent; Sri Lanka.

14. (362) WHITETAILED LAPWING *Vanellus leucurus* M Partridge-: 28 cm. Along jheel edges. Pakistan and N India east to Bangladesh, south to Gujarat and N Madhya Pradesh. Bombay (single record).

15. (363) SOCIABLE LAPWING *Vanellus gregarius* M Partridge±: 33 cm. Dry wasteland, ploughed fields and stubbles. Pakistan and N India east to W Bengal, south through Rajasthan, Gujarat, Maharashtra (irregular) and straggler to Kerala, Sri Lanka.

PLATE 39

1. (364) LAPWING *Vanellus vanellus*
For distribution see Plate 38

2. (366) REDWATTLED LAPWING *Vanellus indicus*
For distribution see Plate 38

3. (365) GREYHEADED LAPWING *Vanellus cinereus*
For distribution see Plate 38

4. (362) WHITETAILED LAPWING *Vanellus leucurus*
For distribution see Plate 38

5. (360) OYSTERCATCHER *Haematopus ostralegus* M Partridge+: 42 cm. Sandy and rocky seashore. Coast line. Subcontinent. Once bred in the Sunderbans (Bangladesh).

6. (432) AVOCET *Recurvirostra avosetta* RM Partridge+: 46 cm. Marshes, lagoons and mudflats, inland and on the sea shore. *Breeding*: N Baluchistan and Kutch. *Winter*: Subcontinent. Sri Lanka (straggler).

7. (434) CRAB PLOVER *Dromas ardeola* M(?) Partridge+: 41 cm. Mudflats and coral reefs. Coast of the subcontinent, Lakshadweep, Andamans, Nicobars and Maldives. Sri Lanka (scarce breeder).

8. (430) BLACKWINGED STILT *Himantopus himantopus* R Partridge-: 25 cm. Jheels, lagoons and reservoirs. Subcontinent; Sri Lanka.

9. (385) WHIMBREL *Numenius phaeopus* M Village hen±: 43 cm. Seashore and mudflats. Subcontinent. Also Lakshadweep, Andamans, Nicobars, Sri Lanka and Maldives.

10. (389) BLACKTAILED GODWIT *Limosa limosa* M Village hen±: male 41 cm; ;female 50 cm. Marshes. Subcontinent; also Lakshadweep, Andamans, Nicobars, Sri Lanka and Maldives.

11. (391) BARTAILED GODWIT *Limosa lapponica* M Village hen±: male 36 cm; female 41 cm. Seashore and mudflats. Coast of Pakistan and W India south to Kerala; also inland (Punjab Salt Range, Bihar); Nicobar I.

12. (403) SNIPEBILLED GODWIT or ASIAN DOWITCHER *Limnodromus semipalmatus* M Partridge+: 34 cm. Seashore and mudflats. Coast of Bangladesh and E India; Chilka Lake (Orissa), NE hill States; Sri Lanka.

13. (388) CURLEW *Numenius arquata* M Village hen±: 58 cm. Seashore and mudflats. Subcontinent including Lakshadweep, Andamans, Nicobars, Sri Lanka and Maldives.

14. (436) STONE CURLEW *Burhinus oedicnemus* R Partridge+: 41 cm. Stony semi-desert. Subcontinent.

15. (437) GREAT STONE PLOVER *Esacus magnirostris* R Village hen±: 51 cm. Rocky river-beds and seacoasts. Subcontinent. Also Andamans and Nicobars; Sri Lanka.

16. (433) IBISBILL *Ibidorhyncha struthersii* R Partridge+: 41 cm. Shingle banks in large river-beds. Himalayas from Gilgit east; *c.* 1700-4400 m (summer), lower elevations (winter).

PLATE 39
1
2
3
4
5
6
7
8
9
10
11
12
13
14
15
16
John M. Dick

PLATE 40
1
2
3
4
5 W
6 W
7 W
5 S
6 S
7 S
9 W
8 W
11
10 W
13 W
14 W
12 W
14 S
15
15 S
16 W
17 W
18 W
16 S
19 W
John H. Dick

PLATE 40

1. (383) LONGBILLED RINGED PLOVER *Charadrius placidus* M Quail+: 23 cm. Shingle banks of large rivers. Bengal, Bangladesh, Assam, Bhutan, Sikkim, Nepal and Andamans.

2. (378) RINGED PLOVER *Charadrius hiaticula* M (or V?) Quail±: 19 cm. Recorded from Karachi, Gilgit, Bharatpur, Delhi, Tamil Nadu, Maldive Is.; frequent in recent years in Sri Lanka.

3. (380) LITTLE RINGED PLOVER *Charadrius dubius* RM Quail-: 17 cm. Shingle beds in rivers, seashore and mudflats. Subcontinent, Andamans; Sri Lanka; Maldives.

4. (381) KENTISH PLOVER *Charadrius alexandrinus* RM Quail-: 17 cm. Sandy shores. Subcontinent. Sri Lanka. *Winter:* Lakshadweep, Nicobars.

5. (384) LESSER SAND PLOVER *Charadrius mongolus* RM Quail±: 19 cm. *Summer* (breeding): Stony or sandy plains near lakes, bogs and streams above *c.* 3400 m. Ladakh, Lahul, Sikkim. *Winter:* Seashore and mudflats. Subcontinent, Andamans, Nicobars, Sri Lanka, Maldives.

6. (374) LARGE SAND PLOVER *Charadrius leschenaultii* M Partridge-: 22 cm, Sandy shores and mudflats. Seaboard of the Subcontinent, Andamans, Nicobars; Sri Lanka; Maldives.

7. (376) CASPIAN SAND PLOVER *Charadrius asiaticus* V Quail±: 19 cm. Seacoast and mudflats. Isolated single specimens each from west coast of the peninsula (Ratnagiri), and Sri Lanka, and a pair from Maldive Is: sight recorded once in the Andamans. Easily mistaken with 384, q.v. in winter plumage.

8. (416) LITTLE STINT *Calidris minuta* M Quail-: 15 cm. Shores, mudflats and marshes. Subcontinent: Ladakh, Kashmir, Nepal, Lakshadweep, Sri Lanka, Maldives.

9. (417) TEMMINCK'S STINT *Calidris temminckii* M Sparrow±: 15 cm. Coastal lagoons, tidal mudflats and inland marshes. Subcontinent: Lakshadweep, Andamans, Maldives.

10. (418) LONGTOED STINT *Calidris subminuta* M Sparrow±: 15 cm. Tidal mudflats and inland marshes. Assam, Bangladesh, W Bengal, Bihar, Nepal, Kerala, Sri Lanka, Maldive and Andaman Is.

11. (419) ASIAN PECTORAL SANDPIPER *Calidris acuminata* V Quail+: 22 cm. Only three definite records: Gilgit, Kashmir and Sri Lanka.

12. (413) EASTERN KNOT *Calidris tenuirostris* M Partridge-: 29 cm. Tidal mudflats. Pakistan, Lakshadweep and Andaman Is; also Assam, Bengal, Kerala and Tamil Nadu. Sri Lanka (new record).

13. (412) KNOT *Calidris canutus* V Quail+: 25 cm. Mudflats and estuaries. One record each from Baluchistan and W Bengal; four from Tamil Nadu; Sri Lanka (frequent).

14. (420) DUNLIN *Calidris alpina* M Quail±: 19 cm. Seashore, mudflats and riverbanks. Pakistan and NW India, on large rivers of the Gangetic system, south to the E & W coasts of the Peninsula; Maldives.

15. (422) CURLEW-SANDPIPER *Calidris testacea* M Quail±: 20 cm. Seashore, mudflats and marshes. Subcontinent, Andamans, Nicobars, Sri Lanka, Maldives.

16. (424) BROADBILLED SANDPIPER *Limicola falcinellus* M Quail- 17 cm. Seashore, mudflats and river banks. Subcontinent, Andamans, Nicobars and Sri Lanka.

17. (428) REDNECKED PHALAROPE *Phalaropus lobatus* M Quail±: M 19 cm, F smaller. Pelagic. Coastal waters, inland ponds and jheels. Pakistan, Gujarat, Tamil Nadu and Sri Lanka. On passage inland (Baluchistan to Bihar) and off the coast of Tamil Nadu.

18. (423) SPOONBILLED SANDPIPER *Eurynorhynchus pygmeus* Rare M or V (?) Quail-: 17 cm. Seashore and mudflats. Bangladesh, W Bengal, Tamil Nadu (Coromandel coast) and Sri Lanka.

19. (414) SANDERLING *Calidris alba* M Quail±: 19 cm. Sandy shores. Coasts of the Subcontinent, Car Nicobar. Nepal (once).

PLATE 41
1
2
3
4 W
5 W
6 W
7 W
8 W
9 W
10 W
11 W
12 W
13
14
15
16
17
18
17
18
John H. Dick

PLATE 41

Waders in flight

1. (378) RINGED PLOVER *Charadrius hiaticula*
   For distribution see Plate 40.
2. (380) LITTLE RINGED PLOVER *Charadrius dubius*
   For distribution see Plate 40.
3. (381) KENTISH PLOVER *Charadrius alexandrinus*
   For distribution see Plate 40.
4. (384) LESSER SAND PLOVER *Charadrius mongolus*
   For distribution see Plate 40.
5. (414) SANDERLING *Calidris alba*
   For distribution see Plate 40.
6. (420) DUNLIN *Calidris alpina*
   For distribution see Plate 40.
7. (422) CURLEW-SANDPIPER *Calidris testacea*
   For distribution see Plate 40.
8. (416) LITTLE STINT *Calidris minuta*
   For distribution see Plate 40.
9. (417) TEMMINCK'S STINT *Calidris temminckii*
   For distribution see Plate 40.
10. (412) KNOT *Calidris canuta*
    For distribution see Plate 40.
11. (413) EASTERN KNOT *Calidris tenuirostris*
    For distribution see Plate 40.
12. (428) REDNECKED PHALAROPE *Phalaropus lobatus*
    For distribution see Plate 40.
13. (409) FANTAIL SNIPE *Gallinago gallinago*
    For distribution see Plate 43.
14. (411) WOODCOCK *Scolopax rusticola*
    For distribution see Plate 43.
15. (406) PINTAIL SNIPE *Gallinago stenura* M Quail+: 27 cm. Marshland, paddy stubble, wet grazing and fallow land. Subcontinent, Andamans, Nicobars, Sri Lanka, Maldives.
16. (429) PAINTED SNIPE *Rostratula benghalensis*
    For distribution see Plate 38.
17. (439) CREAMCOLOURED COURSER *Cursorius cursor*
    For distribution see Plate 38.
18. (440) INDIAN COURSER *Cursorius coromandelicus*
    For distribution see Plate 38.

PLATE 42

Waders

1. (360) OYSTERCATCHER *Haematopus ostralegus*
   For distribution see plate 39.

2. (403) SNIPEBILLED GODWIT or ASIAN DOWITCHER *Limnodromus semipalmatus*
   For distribution see plate 39.

3. (433) IBISBILL *Ibidorhyncha struthersii*
   For distribution see plate 39.

4. (388) CURLEW *Numenius arquata*
   For distribution see plate 39.

5. (389) BLACKTAILED GODWIT *Limosa limosa*
   For distribution see plate 39.

6. (391) BARTAILED GODWIT *Limosa lapponica*
   For distribution see plate 39.

7. (385) WHIMBREL *Numenius phaeopus*
   For distribution see plate 39.

8. (430) BLACKWINGED STILT *Himantopus himantopus*
   For distribution see plate 39.

9. (434) CRAB PLOVER *Dromas ardeola*
   For distribution see plate 39.

10. (432) AVOCET *Recurvirostra avosetta*
    For distribution see plate 39.

11. (436) STONE CURLEW *Burhinus oedicnemus*
    For distribution see plate 39.

12. (437) GREAT STONE PLOVER *Esacus magnirostris*
    For distribution see plate 39.

PLATE 42

1

2 W

3

4

5

6

7

8

9

10

11

12

John H. Dick

PLATE 43
1 W
2 W
3 W
4
5
6
7 W
8 W
9 W
10
11
12 ♀
♂
13
John H. Dick

PLATE 43

1. (393) REDSHANK *Tringa totanus* RM Partridge-: 28 cm. Marshes, margins of lakes, river banks and estuaries. *Breeding:* Kashmir and Ladakh. *Winter:* Subcontinent, Andamans, Nicobars, Sri Lanka, Maldives.

2. (395) MARSH SANDPIPER *Tringa stagnatilis* M Partridge-: 25 cm. Marshes, margins of ponds, inundated fields and mudflats. Subcontinent, Sri Lanka.

3. (396) GREENSHANK *Tringa nebularia* Partridge+: 36 cm. Marshes, margins of ponds, lagoons and mudflats. Subcontinent, Andamans, Nicobars, Sri Lanka, Maldives.

4. (399) SPOTTED GREENSHANK *Tringa guttifer* M Partridge±: 33 cm. Sandbars and mudflats in large rivers and near the sea. Bangladesh and NE India. Nepal (unconfirmed).

5. (398) WOOD SANDPIPER *Tringa glareola* M Quail±: 21 cm. Marshes, flooded paddyfields, mudflats. Subcontinent, Andamans, Sri Lanka, Maldives.

6. (397) GREEN SANDPIPER *Tringa ochropus* RM Quail+: 24 cm. Marshes, small streams and ponds. *Breeding:* Chitral, *c.* 1400 m. *Winter.* Subcontinent, Andamans, Sri Lanka.

7. (400) TEREK SANDPIPER *Tringa terek* M quail+: 24 cm. Seashore, mudflats and lagoons. Seaboard of Subcontinent. Rarely inland; Kashmir, Nepal, Rajasthan and Delhi.

8. (402) TURNSTONE *Arenaria interpres* M Quail±: 22 cm. Rocky seashore. Coasts of Subcontinent, Andamans, Nicobars, Lakshadweep, Sri Lanka, Maldives.

9. (401) COMMON SANDPIPER *Tringa hypoleucos* RM Quail±: 21 cm. Gravel bars in rivers; reservoirs, lagoons and seashore. *Breeding:* N. Baluchistan (?), Kashmir, Ladakh, Garhwal; *c.* 1800-3200 m. *Winter:* Subcontinent, Lakshadweep, Andamans, Nicobars, Sri Lanka, Maldives.

10. (410) JACK SNIPE *Gallinago minima* M Quail±: 21 cm. Marshes. Subcontinent, Andamans, Sri Lanka.

11. (409) FANTAIL SNIPE *Gallinago gallinago* RM Quail+: 27 cm. Marshes. *Breeding:* Kashmir, Ladakh and Garhwal. *Winter:* Subcontinent, Andamans, Sri Lanka, Maldives.

12. (426) RUFF and REEVE *Philomachus pugnax* M Partridge±: 31 cm; F Quail+: 25 cm. Mudflats, marshes and wet paddy stubble. Subcontinent. Also Nepal, Sri Lanka, Maldives.

13. (411) WOODCOCK *Scolopax rusticola* RM Partridge+: 36 cm. Dense shady forest and swampy glades. *Breeding:* NWFP and Kashmir to Arunachal Pradesh; *c.* 2000-4000 m. *Winter:* Nepal, Assam and S Indian hills. Andamans (an isolated record).

PLATE 44

Waders in flight

1. (371) GREY or BLACKBELLIED PLOVER *Pluvialis squatarola*
   For distribution see Plate 38.

2. (442) COLLARED PRATINCOLE *Glareola pratincola pratincola* RM Myna±: 23 cm. Flood-plains near rivers and grazing land by jheels. *Breeding:* Baluchistan and Sind. *Winter:* Allahabad, Ahmedabad, Ratnagiri and Coimbatore. Sri Lanka.

3. (443) (ORIENTAL) COLLARED PRATINCOLE *Glareola pratincola maldivarum*
   For distribution see Plate 38.

4. (444) SMALL INDIAN PRATINCOLE *Glareola lactea*
   For distribution see Plate 38.

5. (372) GOLDEN PLOVER *Pluvialis apricaria* V Partridge-: 27 cm. Muddy shores and wet pastures. Odd specimens identified from Pakistan, Uttar Pradesh and Assam.

6. (402) TURNSTONE *Arenaria interpres*
   For distribution see Plate 43.

7. (373) EASTERN GOLDEN PLOVER *Pluvialis dominica*
   For distribution see Plate 38.

8. (397) GREEN SANDPIPER *Tringa ochropus*
   For distribution see Plate 43.

9. (392) SPOTTED or DUSKY REDSHANK *Tringa erythropus* M Partridge±: 33 cm. Marshes, reservoirs, estuaries. Subcontinent, Sri Lanka.

10. (396) GREENSHANK *Tringa nebularia*
    For distribution see Plate 43.

11. (398) WOOD SANDPIPER *Tringa glareola*
    For distribution see Plate 43.

12. (393) COMMON REDSHANK *Tringa totanus*
    For distribution see Plate 43.

13. (401) COMMON SANDPIPER *Tringa hypoleucos*
    For distribution see Plate 43.

14. (400) TEREK SANDPIPER *Tringa terek*
    For distribution see Plate 43.

15. (426) RUFF and REEVE *Philomachus pugnax*
    For distribution see Plate 43.

PLATE 44
1 ↑
1 ↓
2
3
4
5 ↑
6 ↓
7 ↓
7 ↑
8 ↓
8 ↑
10 W
11 ↓
9 W
11 ↑
12 W
14 W
13 W
15 ♂ W
John M. Dick

PLATE 45
1
2 (tail)
3
4
5
6 w
6 s
7 s
7 imm
8 w
8 imm
9
9 imm
10 imm
11 w
11 imm
12
13 s
13 imm
John H. Dick

PLATE 45

1. (445) ANTARCTIC SKUA *Catharacta skua antarctica* V Duck±: 53-61 cm. Five records, Sri Lanka; one possible record, Maldive Is.

2. (447) POMATORHINE SKUA or JAEGER *Stercorarius pomarinus* V Duck-: 53 cm. Sight records from Pakistan (off Karachi); Sri Lanka (quite a few in recent years).

3. (484) INDIAN SKIMMER *Rhynchops albicollis* R Crow±: 40 cm. Large rivers. The Indus, Ganges and Brahmaputra river systems, Nepal terai and the large rivers of E and peninsular India. Vagrant elsewhere.

4. (446a) ANTARCTIC SKUA (MACCORMICK'S SKUA) *Catharacta skua maccormicki* V Duck±: 53-61 cm. One record each from Sri Lanka and Karnataka.

5. (448) PARASITIC or RICHARDSON'S SKUA or JAEGER *Stercorarius parasiticus* V (or M?) Duck-: 48 cm. Pakistan coast and the Arabian Sea. Sri Lanka.

6. (453) GREAT BLACKHEADED GULL *Larus ichthyaetus* M Duck+: 66-72 cm. Sea coast and large rivers and lakes. Subcontinent; Sri Lanka. Occasionally inland: Nepal, Gujarat, Rajasthan.

7. (449) SOOTY GULL *Larus hemprichii* M Duck-: 48 cm. Sea coast. Pakistan; Bombay (straggler).

8. (450) HERRING GULL *Larus argentatus* M Duck±: 60 cm. Sea coast and inland lakes. Coasts of Pakistan and India; the Ganges river. One record each from Nepal and Calcutta. Chitral (passage).

9. (452) LESSER BLACKBACKED GULL *Larus fuscus* M Duck±: 60 cm. Sea coast. Pakistan and India. Nepal (once); Sri Lanka.

10. (457) LITTLE GULL *Larus minutus* V Pigeon-: 25-30 cm. Recorded from Ladakh and Great Rann of Kutch.

11. (455) BLACKHEADED GULL *Larus ridibundus* M Crow±: 43 cm. Sea coast, estuaries, etc. Subcontinent; Nepal; Sri Lanka; Maldives (vagrant).

12. (456) SLENDERBILLED GULL *Larus genei* RM Crow±: 43 cm. Jheels and sea coast. *Breeding:* Baluchistan. *Winter:* Sind, Gujarat, Bombay, Tamil Nadu and Nepal terai. Sri Lanka (sight).

13. (454) BROWNHEADED GULL *Larus brunnicephalus* RM Crow+: 46 cm. Sea coast, lagoons, backwaters, estuaries and large rivers. *Breeding:* Ladakh. *Winter:* Subcontinent. Also Nepal; Sri Lanka.

PLATE 46

1. (459a) BLACK TERN *Chlidonias niger* V Pigeon-: 23 cm. Sight records, Delhi, Gujarat, Andhra; Moskow ringed individual recovered at Pt. Calimere, Tamil Nadu.

2. (459) WHITEWINGED BLACK TERN *Chlidonias leucopterus* M Pigeon-: 23 cm. Marshes and rivers. Nepal, Assam, Bangladesh and Sri Lanka. Irregularly in the rest of the Indian peninsula, Andamans; Maldives.

3. (458) WHISKERED TERN *Chlidonias hybridus* RM Pigeon-: 25 cm. Lakes, marshes, coastal lagoons and mudflats. *Breeding:* Kashmir, N India in the Gangetic plain, Assam and Bangladesh. *Winter:* Subcontinent in Pakistan, E Nepal, and Sri Lanka.

4. (475) LITTLE TERN *Sterna albifrons* R Myna±: 23 cm. Rivers, marshes and estuaries. Pakistan, Gujarat, coast of Maharashtra, large rivers of N India, Nepal, Bangladesh and Sri Lanka.

5. (483) INDIAN OCEAN WHITE TERN or FAIRY TERN *Gygis alba* R Pigeon-: 27-33 cm. Pelagic. Addu Atoll, Maldives. Straggler to the Bay of Bengal.

6. (466) ROSEATE or ROSY TERN *Sterna dougalli* R Crow-: 38 cm. Sea coast. Islets off the coast of S India and Sri Lanka; the Maldive (?) and Andaman Is.

7. (467) WHITECHEEKED TERN *Sterna repressa* R Crow-: 35 cm. Pelagic. *Breeding:* Vengurla Rocks off the Maharashtra coast. Coasts of Pakistan, W India and Lakshadweep Is.

8. (464) COMMON TERN *Sterna hirundo* RM Crow-: 36 cm. Large rivers and estuaries. *Breeding:* Ladakh, Sri Lanka. *Winter:* Subcontinent; Maldives.

9. (463) INDIAN RIVER TERN *Sterna aurantia* R Crow±: 38-46 cm. Large rivers and reservoirs. Subcontinent; absent in Sri Lanka.

10. (462) CASPIAN TERN *Hydroprogne caspia* RM Crow+: 51 cm. Sea coast and large lakes. *Breeding:* Baluchistan and Sri Lanka. *Winter:* Subcontinent.

11. (478) LARGE CRESTED TERN *Sterna bergii* R Kite-: 53 cm. Pelagic. *Breeding:* Islets off the coasts of Pakistan, peninsular India, W. Bengal, Bangladesh and Sri Lanka; Lakshadweep and probably Maldives. Post-breeding dispersal: entire seaboard.

12. (468) BLACKNAPED TERN *Sterna sumatrana* R Crow-: 33 cm. Pelagic. Andaman, Nicobar and Maldive Is.

13. (470) BLACKBELLIED TERN *Sterna acuticauda* R Crow-: 33 cm. Rivers. Subcontinent east of the Indus; absent in Sri Lanka.

14. (479) INDIAN LESSER CRESTED TERN *Sterna bengalensis* R Crow±: 43 cm. Offshore waters. Subcontinent. Presumably breeding on the Makran coast, Rameshwaram, Lakshadweep and Maldives. Also occurs in Andamans, Nicobars and Sri Lanka.

15. (460) GULLBILLED TERN *Gelochelidon nilotica* RM Crow-: 38 cm. Jheels, rivers and coastal mudflats. *Breeding:* Pakistan, W Bengal and Bangladesh. *Winter:* Subcontinent, Andamans (once); Sri Lanka; Maldives.

16. (474) SOOTY TERN *Sterna fuscata* R Crow±: 43 cm. Pelagic. Lakshadweep and probably also Maldive and Andaman Is. Has strayed to Tamil Nadu, Bihar and Assam.

17. (481) NODDY TERN *Anous stolidus* R Crow-: 41 cm. Pelagic. *Breeding:* Lakshadweep, Andaman, Nicobar, and possibly also the Maldives. Straggler to Pakistan and Sri Lanka.

18. (480) SANDWICH TERN *Sterna sandvicensis* M Crow+: 44 cm. Sea coast. Pakistan, Gujarat and Sri Lanka.

19. (471) BROWNWINGED TERN *Sterna anaethetus* R Crow-: 37 cm. Pelagic. *Breeding:* Islands off the Maharashtra coast, Makran coast (?), Lakshadweep and Maldive Is. Post-breeding dispersal chiefly over the Arabian Sea to the coasts of Pakistan, W India and Sri Lanka. Andamans (two records).

1 S
1 W
2 S
2 W
3 S
3 W
4 W
4 S
5
6 S
7 S
8 S
8 W
9 S
10 W
11 br +
11 br –
12
13
14 S
15 S
16 S
17
18 W
19 S
John H. Dick

PLATE 47
1♂
♀
♂
1♀
1♂
2♀
2♂
4♂
♀
5♂
3♂
4♀
6♂
5♀
4♂
5♂
6♀
7♂
8♂
6♂
7♀
8♀
7♂
8♂
7♂ br –
9♀
9
9♂
John H. Dick

PLATE 47

1. (489) IMPERIAL SANDGROUSE *Pterocles orientalis* R Pigeon+: 39 cm. Sandy semi-desert. Pakistan and NW India; vagrant to Nepal, Uttar Pradesh, Madhya Pradesh and Karnataka. *Breeding:* Baluchistan and Sind.

2. (490) CORONETTED SANDGROUSE *Pterocles coronatus* R Pigeon-: 28 cm. Stony semi-desert. Pakistan west of the Indus.

3. (491) CLOSEBARRED SANDGROUSE *Pterocles indicus arabicus* R Pigeon-: 27 cm. Stony semi-desert. Pakistan west of the Indus.

4. (487) INDIAN SANDGROUSE *Pterocles exustus* R Pigeon±: 28 cm. Stony semi-desert and dry fallow land. Pakistan, and India east to Bengal, south to Tamil Nadu.

5. (492) PAINTED SANDGROUSE *Pterocles indicus indicus* R Pigeon-: 28 cm. Bare, stony plateau country. Pakistan east of the Indus, and India east to Bihar and Orissa, south to Karnataka and Tamil Nadu.

6. (488) SPOTTED SANDGROUSE *Pterocles senegallus* R Pigeon+: 36 cm. Open semi-desert. Pakistan and NW India. *Breeding:* Baluchistan.

7. (486) LARGE PINTAIL SANDGROUSE *Pterocles alchata* (R?) M Pigeon+: 38 cm. Sandy or stony semi-desert. Pakistan, Haryana, Delhi, Rajasthan and Gujarat.

8. (485a) PALLAS'S SANDGROUSE *Syrrhaptes paradoxus* V Crow+: 48 cm. Single record, Rajasthan.

9. (485) TIBETAN SANDGROUSE *Syrrhaptes tibetanus* R Crow+: 48 cm. Stony, semi-desert steppes. Ladakh and Himachal Pradesh in Tibetan facies; *c.* 4200-5400 cm.

PLATE 48

1. (516) BLUE ROCK PIGEON *Columba livia* R 33 cm. Cliffs and gorges, and around habitations and cultivation. Subcontinent, Nicobars (introduced); Sri Lanka.

2. (EL) STOCK PIGEON *Columba oenas* Europe, Mediterranean, W Asia.

3. (518) EASTERN STOCK PIGEON *Columba eversmanni* M Pigeon-: 30 cm. Groves of trees in cultivated country. Pakistan, and NW India east to Bihar, south to Rajasthan and Gwalior.

4. (515) HILL PIGEON *Columba rupestris* R Pigeon±: 33 cm. Cliffs and gorges in high plateau country. Himalayas, *c.* 3000-5500 m; lower in winter.

5. (513) SNOW PIGEON *Columba leuconota* R Pigeon±: 34 cm. Cliffs and steep gorges. Himalayas, *c.* 3000-4500 m; lower in winter.

6. (519) WOOD PIGEON *Columba palumbus* R Pigeon+: 43 cm. Forests of oak and conifers. Baluchistan, Salt Range and W Himalayas. Seasonally between *c.* 800 and 3500 m. C Nepal (winter).

7. (521) NILGIRI WOOD PIGEON *Columba elphinstonii* R Pigeon+: 42 cm. Moist-evergreen forest and sholas. W Ghats and associated hills from Bombay south through Kerala.

8. (523) ASHY WOOD PIGEON *Columba pulchricollis* R Pigeon+: 36 cm. Mixed deciduous and evergreen forest. E Himalayas *c.* 1200-3200 m; NE hill states.

9. (506) GREEN IMPERIAL PIGEON *Ducula aenea* R Crow+: 43 cm. Evergreen and moist-deciduous forest. Subcontinent; also Aandamans, Nicobars, Bangladesh; Sri Lanka.

10. (509) PIED IMPERIAL PIGEON *Ducula bicolor* R Pigeon+: 41 cm. Forest, especially mangroves. Andaman and Nicobar Is.

11. (524) PURPLE WOOD PIGEON *Columba punicea* R Pigeon+: 36 cm. Well-wooded ravines with dense evergreen undergrowth or where interspersed with scrub jungle and cultivation. Bihar, E Madhya Pradesh and Orissa east to Assam, Nagaland, Manipur and Bangladesh (Chittagong region).

12. (522) CEYLON WOOD PIGEON *Columba torringtoni* R Pigeon+: 36 cm. Evergreen and moist-deciduous forest. Sri Lanka.

13. (525) ANDAMAN WOOD PIGEON *Columba palumboides* R Pigeon+: 41 cm. Evergreen forest. Andaman and Nicobar Is.

14. (520) SPECKLED WOOD PIGEON *Columba hodgsonii* R Pigeon+: 38 cm. Tall evergreen and semi-evergreen forest. Himalayas *c.* 1800-4000 m; NE hill states.

15. (510) IMPERIAL PIGEON *Ducula badia* R Crow+: 51 cm. Tall evergreen forest. E Himalayas; NE hill states, up to *c.* 2300 m; W Ghats complex.

PLATE 48
1
2
3
4
1
5
5
6
6
7
10
8
9
10
11
12
13
14
15
15
John M. Dick

PLATE 49
John H. Dick

PLATE 49

1. (494) WEDGETAILED GREEN PIGEON *Treron sphenura* R Pigeon±: 33 cm. Broad-leafed forest. Lower Himalayas from Kashmir to Bhutan and Arunachal Pradesh, and hills south of the Brahmaputra; up to *c.* 2500 m.

2. (503) YELLOWLEGGED GREEN PIGEON *Treron phoenicoptera* R Pigeon±: 33 cm. Dry- and moist-deciduous biotope. Pakistan (Punjab) and N from Himalayan foothills south through the peninsula; Bangladesh; Sri Lanka.

3. (496) POMPADOUR or GREYFRONTED GREEN PIGEON *Treron pompadora* R Pigeon-: 28 cm. Tropical evergreen and moist-deciduous forest. Nepal, W Bengal, NE India and Bangladesh; W Ghats complex south to Kerala, Tamil Nadu (Nlgiris); Sri Lanka; Andaman and Nicobar Is.

4. (495) THICKBILLED GREEN PIGEON *Treron curvirostra* R Pigeon-: 27 cm. Well-wooded country up to *c.* 1500 m. Lower Himalayas from W Nepal (up to 2750 m) east to Bhutan and Arunachal Pradesh; NE hill states to Bangladesh and Manipur.

5. (493) PINTAILED GREEN PIGEON *Treron apicauda* R Pigeon+: M 42 cm including tail of *c.* 21 cm; F tail shorter *c.* 18 cm. Tall forest. Lower Himalayas from Kumaon to Bhutan and Arunachal Pradesh; up to *c.* 1800 m.

6. (501) ORANGEBREASTED GREEN PIGEON *Treron bicincta* R Pigeon-: 29 cm. Evergreen and moist-deciduous forest. Lower Himalayas (up to 1500 m) from the Uttar Pradesh terai and Nepal east to Bhutan and Arunachal Pradesh, Manipur, W Bengal and Bangladesh; E Ghats, and W Ghats from Belgaum southwards, Tamil Nadu (Pt. Calimere), Sri Lanka.

7. (544b) NICOBAR PIGEON *Caloenas nicobarica* R Pigeon+: 41 cm. Heavy evergreen forest. Nicobar Is., South Sentinel Is. (off Little Andaman) and Coco Is.

8. (526) BARTAILED CUCKOO-DOVE *Macropygia unchall* R Pigeon+: 41 cm. Dense evergreen forest and secondary jungle. Himalayas from Kashmir to Arunachal, and south through NE hill states to Meghalaya and Manipur.

9. (542) EMERALD or BRONZEWINGED DOVE *Chalcophaps indica* R Myna+: 27 cm. Evergreen and moist-deciduous jungle. Subcontinent; Sri Lanka; Andaman and Nicobar Is.

10. (527) ANDAMAN CUCKOO-DOVE *Macropygia rufipennis* R Pigeon+: 41 cm. Dense evergreen forest and secondary jungle. Andaman and Nicobar Is.

11. (535) RED TURTLE DOVE *Streptopelia tranquebarica* R Myna±: 23 cm. Deciduous scrub jungle and cultivation. Subcontinent; normally under 800 m in the Himalayas. Andamans; Sri Lanka (three records).

12. (534) INDIAN RING DOVE *Streptopelia decaocto* R Pigeon±: 32 cm. Dry-deciduous plains country with light jungle and cultivation. Subcontinent; Sri Lanka.

13. (541) LITTLE BROWN or SENEGAL DOVE *Streptopelia senegalensis* R Myna+: 27 cm. Dry-deciduous biotope. Subcontinent east to Nepal, Bangladesh and Tripura; not recorded in Sri Lanka.

14. (537) SPOTTED DOVE *Streptopelia chinensis* R Pigeon-: 30 cm. Dry- and moist-deciduous biotope — better wooded moister facies than Ring or Little Brown. Subcontinent; Bangladesh; Sri Lanka.

15. (529) TURTLE DOVE *Streptopelia turtur* V Pigeon-: 28 cm. Open cultivated country. Pakistan.

16. (530) RUFOUS TURTLE DOVE *Streptopelia orientalis* RM Pigeon±: 33 cm. Moist-deciduous biotope, mixed forest and bamboo jungle. Subcontinent, to *c.* 4000 m in the Himalayas. Sri Lanka (absent).

1
2 ♀
3 ♂
2 ♂
4
5 ♂
5 ♀
6 ♂
9 ♂
7 ♂
8 ♂
9 ♀
13
10 ♂
11 ♂
14
14
12 ♂
John H. Dick

PLATE 50

1. (553) NICOBAR PARAKEET *Psittacula caniceps* R Pigeon±: overall 61 cm. High evergreen forest. Nicobar Is.

2. (554) LORD DERBY'S PARAKEET *Psittacula derbyana* Summer M Myna+: overall 46 cm. Conifer forest and cultivation. N Arunachal Pradesh, *c.* 2700-3500 m.

3. (546) ALEXANDRINE PARAKEET *Psittacula eupatria* R Pigeon+: 53 cm. Dry- and moist-deciduous biotope. Subcontinent, Andamans, Nicobars; Bangladesh; Sri Lanka.

4. (563) EASTERN SLATYHEADED PARAKEET *Psittacula finschii* R Myna-: overall 36 cm. Forest and cultivation. SE Bhutan and Arunachal Pradesh south through NE hill states to Bangladesh; up to *c.* 2100 m.

5. (558) BLOSSOMHEADED PARAKEET *Psittacula cyanocephala* R Myna-: overall 36 cm. Lowlands and hills to *c.* 1500 m. Lower Himalayas of Pakistan east to Bhutan, south throughout the Peninsula; Sri Lanka.

6. (562) SLATYHEADED PARAKEET *Psittacula himalayana* R Myna+: overall 41 cm. Well-wooded hillsides and valleys. Himalayas, seasonally between *c.* 600 and 2500 m.

7. (559) EASTERN BLOSSOMHEADED PARAKEET *Psittacula roseata* R Myna-: overall 36 cm. Light forest and cultivation. Sikkim duars and east through NE hill states to Bangladesh.

8. (555) REDCHEEKED PARAKEET *Psittacula longicauda* R Myna±: overall 46-48 cm. Evergreen forest and cultivation. Andaman and Nicobar Is.

9. (550) ROSERINGED PARAKEET *Psittacula krameri* R Myna+: overall 42 cm. Moist- and dry-deciduous biotope, thin jungle, orchards, cultivation and human environments. Subcontinent; Bangladesh; Sri Lanka.

10. (564) BLUEWINGED PARAKEET *Psittacula columboides* R Myna±: overall 38 cm. Evergreen and moist-deciduous biotope. W Ghats from the Thane district (*c.* 19°N lat.) south through Kerala and associated hills of Karnataka and Tamil Nadu; *c.* 500-1000 m.

11. (551) REDBREASTED PARAKEET *Psittacula alexandri* R Pigeon+: overall 38 cm. Moist-deciduous biotope, thin secondary jungle. Lower Himalayas from Kumaon east to Bhutan and Arunachal Pradesh up to *c.* 1500 m, NE hill states; Andaman Is. Bangladesh.

12. (565) LAYARD'S PARAKEET *Psittacula calthropae* Endemic Myna-: overall 31 cm. Sri Lanka in the low country and hill forest up to *c.* 1800 m.

13. (568) CEYLON LORIKEET *Loriculus beryllinus* Endemic Sparrow±: 14 cm. Evergreen and moist-deciduous biotope. Sri Lanka.

14. (566) INDIAN LORIKEET *Loriculus vernalis* R Sparrow±: 14 cm. Evergreen and moist-deciduous forest. E Himalayan foothills; NE hill states; the Indian peninsula in E Ghats, and W Ghats from Khandala (*c.* 18°N), south through Karnataka, Kerala and Tamil Nadu; Andaman and Nicobar Is.

PLATE 51

1. (572) LARGE HAWK-CUCKOO *Cuculus sparverioides* RM Crow-: 38 cm. Dense forest. Himalayas, *c.* 900-3200 m; NE hill states. Karnataka, Kerala, Tamil Nadu and Andhra (E. Ghats) [winter].

2. (575) HODGSON'S HAWK-CUCKOO *Cuculus fugax* R Pigeon-: 29 cm. Deciduous and evergreen forest. Lower E Himalayas; NE hill states; Bangladesh; *c.* 600-1800 m.

3. (573) COMMON HAWK-CUCKOO or BRAINFEVER BIRD *Cuculus varius* R Pigeon±: 34 cm. Deciduous and semi-evergreen forest, groves and gardens. Subcontinent; Sri Lanka.

4. (586) EMERALD CUCKOO *Chalcites maculatus* R or M? Sparrow+: 18 cm. Evergreen forest. Apparently summer visitor to Himalayas. From Garhwal to Bhutan and Arunachal Pradesh; NE hill states. *Winter:* Andaman and Nicobar Is.; Tamil Nadu, C India and Sri Lanka (vagrant).

5. (585) RUFOUSBELLIED PLAINTIVE CUCKOO *Cacomantis merulinus* R Myna±: 23 cm. Well-wooded country. E Himalayas (up to *c.* 2000 m); NE hill states; Bangladesh; Assam; W Bengal. Bihar, Madhya Pradesh, Andhra (winter).

6. (582) INDIAN BAYBANDED CUCKOO *Cacomantis sonneratti* R Myna±: 24 cm. Deciduous and evergreen wooded country. Subcontinent; up to *c.* 2400 m in Himalayas; Sri Lanka.

7. (587) VIOLET CUCKOO *Chalcites xanthorhynchus* R Sparrow+: 17 cm. Secondary evergreen forest. NE hill states; Bangladesh; Andaman and Nicobar Is. Up to *c.* 1500 m. Tamil Nadu (vagrant).

8. (584) INDIAN PLAINTIVE CUCKOO *Cacomantis passerinus* R Myna±: 23 cm. Open forest, well-wooded country and cultivation. Subcontinent; Sri Lanka. Absent in Pakistan and drier parts of NW India.

9. (581) SMALL CUCKOO *Cuculus poliocephalus* RM Myna+: 26 cm. *Summer:* Himalayas; NE hill states; Bangladesh. Seasonally through the Peninsula and Sri Lanka.

10. (576) INDIAN CUCKOO *Cuculus micropterus* RM Pigeon±: 33 cm. Deciduous and evergreen biotope. Subcontinent, Andamans, Nicobars; Sri Lanka.

11. (578) THE CUCKOO *Cuculus canorus* RM Pigeon±: 33 cm. *Summer:* Himalyas; NE hill states; Bangladesh. Seasonally most hilly regions of the Subcontinent. Andamans; Nicobars; Sri Lanka (winter).

1
2
imm
3
4♂
♀
imm
5
6
7♂
♀
imm
8
♀
11♂
9
imm?
10
John H. Dick

PLATE 52

PLATE 52

1. (595) SMALL GREENBILLED MALKOHA *Rhopodytes viridirostris* R Crow-: 39 cm. Deciduous and semi-evergreen scrub and secondary jungle. Peninsular India from Gujarat, Maharashtra and Orissa south to Kanyakumari; Rameshwaram I.; Sri Lanka.

2. (571) PIED CRESTED CUCKOO *Clamator jacobinus* RM Myna±: 33 cm. Dry- and moist-deciduous lightly wooded country. Seasonally entire Subcontinent; Sri Lanka.

3. (569) REDWINGED CRESTED CUCKOO *Clamator coromandus* R Crow+: 47 cm. Evergreen and moist-deciduous biotope: foothills forest and scrub jungle. Himalayas from Garhwal east to Bhutan and Arunachal Pradesh; NE hill states; Bangladesh. *Winter:* S India (rare); Sri Lanka.

4. (593) LARGE GREENBILLED MALKOHA *Rhopodytes tristis* R Pigeon± with tail *c.* 38 cm; overall length *c.* 51 cm. Dense shrubbery and thickets in evergreen and moist-deciduous biotopes. Lower Himalayas from Garhwal to Arunachal Pradesh; NE hill states; Bangladesh; Bihar, E Madhya Pradesh, E Ghats and N Orissa; up to *c.* 1800 m.

5. (598) SIRKEER CUCKOO *Taccocua leschenaultii* R Crow±: 42-44 cm. Dry-deciduous secondary forest and scrub-and-bush jungle, etc. Subcontinent; Sri Lanka.

6. (605) LESSER COUCAL *Centropus toulou* R Crow-: 33 cm. Scrub jungle and tall grassland. Himalayan submontane terai; Uttar Pradesh east to Nepal; Bhutan; E India; Bangladesh; Bihar and Orissa; also SW India from N Kanara through W Karnataka and Kerala.

7. (599) REDFACED MALKOHA *Phaenicophaeus pyrrhocephalus* R Crow+: 46 cm. Tall evergreen forest. S Kerala, S Tamil Nadu, and Sri Lanka. Rare and endangered.

8. (600) CROW-PHEASANT or COUCAL *Centropus sinensis* R Crow+: 48 cm. Deciduous scrub jungle and gardens. Subcontinent; Bangladesh; Sri Lanka.

9. (604) CEYLON COUCAL *Centropus chlororhynchus* Endemic Crow±: 43 cm. Humid forest. Sri Lanka. Declining in numbers owing to loss of habitat.

10. (603) ANDAMAN CROW-PHEASANT *Centropus sinensis andamanensis* Endemic Crow+: 48 cm. Evergreen forest and mangroves. Andaman, Coco, and possibly Nicobar Is.

PLATE 53

1. (627) EAGLE-OWL, GREAT HORNED OWL *Bubo bubo* R Kite-: 56 cm. Rocky hills with outscoured steep-sided ravines, and well-wooded country. Subcontinent; no record from Bhutan and Bangladesh eastwards.

2. (628) FOREST EAGLE-OWL *Bubo nipalensis* R Kite±: 63 cm. Dense evergreen and moist-deciduous forest up to *c.* 2100 m. Himalayan submontane tracts from Kumaon to Bhutan and Arunachal Pradesh; NE hill states; Bangladesh; also the hills of W Ghats complex in S India, east to the Shevaroy Hills; Sri Lanka.

3. (630) DUSKY HORNED OWL *Bubo coromandus* R Kite±: 58 cm. Wooded, well-watered country with groves of large leafy trees. Most of the Subcontinent south of Himalayas.

4. (659) BROWN WOOD OWL *Strix leptogrammica* R Kite-: 47-53 cm. Dense semi-evergreen and moist-deciduous forest. Himalayas from Pakistan (Punjab) east to Arunachal Pradesh (750-2450 m); NE hill states; rest of the Subcontinent; Sri Lanka.

5. (657) MOTTLED WOOD OWL *Strix ocellata* R Kite±: 48 cm. Wooded plains country: groves of densely foliaged ancient trees in village environs, etc. Pakistan and most of northern and peninsular India.

6. (663) LONGEARED OWL *Asio otus* RM Pigeon+: 37 cm. Hill forest (summer); grassy lowland jungle (winter). *Breeding:* Baluchistan and Kashmir, up to *c.* 2000 m. *Winter:* Pakistan and N India.

7. (664) SHORTEARED OWL *Asio flammeus* M Pigeon+: 38 cm. Open undulating grasslands, scrub and marshes; plains and hills up to *c.* 1400 m. Subcontinent. Occasional in Sri Lanka and Maldives.

8. (631) BROWN FISH OWL *Bubo zeylonensis* R Kite-: 56 cm. Well-wooded and well-watered country. Subcontinent.

9. (662) HIMALAYAN WOOD OWL *Strix aluco* R Crow+: 45-47 cm. Forest of oak, pine and fir. Baluchistan to Chitral, and east through Himalayas (*c.* 1200-4000 m), and south through NE hill states.

10. (654) HUME'S WOOD OWL *Strix butleri* V Pigeon+: 36 cm. Oases and nearby rocks or ruins. Single record, Ormara, Makran coast.

11. (633) TAWNY FISH OWL *Bubo flavipes* R Kite±: 61 cm. Forest near streams. Outer Himalayas (up to *c.* 1500 m) from Kashmir to Bhutan, Arunachal Pradesh (?); NE hill states; Bangladesh.

PLATE 53
1
2
3
4
5
6
7
7
8
9
10
11
John H. Dick

PLATE 54

PLATE 54

1. (614) STRIATED or PALLID SCOPS OWL *Otus brucei* R Myna±: 22 cm. Arid semi-desert and stony foothills country. Pakistan. Also obtained sporadically from Bombay, Poona, Thane, Ahmednagar and Ratnagiri (vagrants?).

2. (612) SPOTTED SCOPS OWL *Otus spilocephalus* R Myna±: 18-20 cm. Dense evergreen forest. Himalayan foothills and duars, and south through NE hill states to Bangladesh (Chittagong region); up to *c.* 2700 m.

3. (617) SCOPS OWL *Otus scops* RM Myna-: 19 cm, Evergreen and deciduous forest. Subcontinent, Andamans, Nicobars; Bangladesh; Sri Lanka.

4. (623) COLLARED SCOPS OWL *Otus bakkamoena* R Myna±:: 23-25 cm. Well-wooded country. Subcontinent; Bangladesh; Sri Lanka.

5. (638) Chestnutbacked ssp. *castanonotum* of 636. Myna-: 19 cm. Moist forest. Sri Lanka.

6. (639) BARRED OWLET *Glaucidium cuculoides* R Myna±: 23 cm. Broadleafed or conifer forest and subtropical evergreen jungle. Outer Himalayas; NE hill states; Bangladesh; up to *c.* 2700 m.

7. (635) COLLARED PIGMY OWLET *Glaucidium brodiei* R Quail±: 17 cm. Deciduous or conifer forest. Himalayas; NE hill states; Bangladesh; between *c.* 900 and 3200 m.

8. (636) JUNGLE OWLET *Glaucidium radiatum* R Myna±: 20 cm. Moist-deciduous forest and mixed secondary jungle. Subcontinent; Sri Lanka.

9. (665) TENGMALM'S OWL or BOREAL OWL *Aegolius funereus* R Myna+: 25 cm. Subalpine juniper forest. Single record of a breeding female and a feathered juvenile from Lahul (Himachal Pradesh).

10. (648, 649) LITTLE OWL *Athene noctua* R Myna±: 23 cm. Semi-desert, rocky country up to *c.* 4600 m. Pakistan (Baluchistan north to NWFP); Baltistan, Ladakh, Nepal.

11. (653) FOREST SPOTTED OWLET *Athene blewitti* Extinct? Myna±: 23 cm. Moist-deciduous jungle and groves of wild mango, especially near streams. From Khandesh and Surat Dangs east along the Satpuras to Madhya Pradesh and Orissa. Less than a dozen specimens known, the latest in 1914.

12. (652) SPOTTED OWLET *Athene brama* R Myna±: 21 cm. Ruins, mango topes and village groves of ancient trees, etc. Subcontinent. Absent in Sri Lanka and Andaman and Nicobar Is.

13. (645) Ssp. *obscura* of 642. Andaman and Nicobar Is.

14. (642) BROWN HAWK-OWL *Ninox scutulata* R Pigeon±: 32 cm. Well-wooded country. Most of the Subcontinent; Bangladesh; Sri Lanka. Not in Pakistan.

15. (609) BAY OWL *Phodilus badius* R (rare) Pigeon-: 29 cm. Dense evergreen submontane forest. Nepal, Sikkim, Bhutan(?), Assam, Meghalaya, Nagaland and Manipur; Sri Lanka; Kerala (a single example in Nelliampathy Hills).

16. (606) BARN OWL *Tyto alba* R Crow-: 36 cm. Buildings and ruins. Subcontinent; Andaman Is. Sri Lanka.

17. (608) GRASS OWL *Tyto capensis* R Crow-36 cm. Tall grass jungle and grassy depressions. Himalayan submontane tracts: Uttar Pradesh east to Assam and south through Bangladesh, Bihar, Orissa and Madhya Pradesh; also S Indian hills; up to *c.* 1800 m.

PLATE 55

1. (666) CEYLON FROGMOUTH *Batrachostomus moniliger* R Myna±: 23 cm. Dense evergreen forest and secondary jungle. W Ghats from N Kanara south; Sri Lanka.

2. (667) HODGSON'S FROGMOUTH *Batrachostomus hodgsoni* R Myna+: 27 cm. Subtropical evergreen foothills forest. E Himalayas; NE hill states; Bangladesh; up to *c.* 1800 m.

3. (674) SYKES'S NIGHTJAR *Caprimulgus mahrattensis* RM Myna±: 23 cm. Stony wastes and open hillsides; also sandy sarkan grass areas. Sind (Pakistan), Kutch, Punjab and Haryana. *Winter:* Kanpur, Madhya Pradesh, Maharashtra and south in the Peninsula to Belgaum.

4. (673) EUROPEAN NIGHTJAR *Caprimulgus europaeus* M Myna+: 25 cm. Open forest or bush-covered hillsides. *Summer:* N Baluchistan, Rawalpindi and the Salt Range in Pakistan. Also Gilgit, Baltistan (?), Kashmir (?) and Garhwal in India. *Winter:* Makran, Sind and Kutch. Straggler to Rajasthan, Maharashtra, Madhya Pradesh and Uttar Pradesh.

5. (673a) EGYPTIAN NIGHTJAR *Caprimulgus aegyptius* M or V Myna+: 25 cm. Semi-desert. Probably a summer visitor to Baluchistan.

6. (671) INDIAN JUNGLE NIGHTJAR *Caprimulgus indicus* RM Pigeon-: 29 cm. Dry- and moist-deciduous jungle. Himalayas; NE hill states; Bangladesh; rest of Subcontinent and Sri Lanka. Andamans (a migrating individual at sea).

7. (682) FRANKLIN'S or ALLIED NIGHTJAR *Caprimulgus affinis* R Myna+: 25 cm. Scrubby hillsides, light deciduous forest and grass jungle. Subcontinent from the Punjab Salt Range, Rajasthan and Gujarat eastwards. Absent in Sri Lanka.

8. (680) COMMON INDIAN NIGHTJAR *Caprimulgus asiaticus* R Myna±: 24 cm. Dry-deciduous scrub-jungle. Subcontinent east of the Indus valley; Sri Lanka.

9. (675) LONGTAILED NIGHTJAR *Caprimulgus macrurus* RM Pigeon-: 28 cm. Moist-deciduous biotope: mixed bamboo and secondary scrub forest. Himalayan foothills (up to *c.* 2400 m) from Himachal Pradesh through Nepal and Bhutan to Arunachal Pradesh; NE hill states; Bangladesh and practically the entire Subcontinent; Sri Lanka; Andaman Is.

10. (669) GREAT EARED NIGHTJAR *Eurostopodus macrotis* R Crow-: 41 cm. Evergreen and moist-deciduous forest and bush jungle. Sikkim (?); Assam; NE hill states; Bangladesh; up to *c.* 1000 m; Kerala.

♂
1 ♀
♂
2 ♀
3
5 ↑
4
5
6
7
8
9 ↑
9
10

John H. Dick

PLATE 56

PLATE 56

1. (692) WHITERUMPED SPINETAIL *Chaetura sylvatica* R Sparrow-: 11 cm. Evergreen and moist-deciduous biotope. Lower Himalayas in Garhwal, Nepal, Sikkim; Assam, the Indian peninsula from Chota Nagpur and W Bengal south through Orissa, Madhya Pradesh, E Maharashtra and N Andhra Pradesh; also W Ghats complex from N Kanara south through Kerala.

2. (683) HIMALAYAN SWIFTLET *Collocalia brevirostris* R Sparrow-: 14 cm. Open areas near forest. Himalayas; NE hill states; Bangladesh; up to *c.* 3600 m. Andamans (winter).

3. (687) WHITEBELLIED SWIFTLET *Collocalia esculenta* R Sparrow-: 10 cm. Buildings. Andaman and Nicobar Is.

4. (703) HOUSE SWIFT *Apus affinis* RM Sparrow±: 15 cm. Around buildings and cliffs. From NWFP east through Punjab, Kashmir and Uttar Pradesh to Assam and Bangladesh; south throughout the Peninsula. Sri Lanka.

5. (697) PALLID SWIFT *Apus pallidus* M Bulbul-: 17 cm. A desert form. The Makran coast and east to Sind (winter).

6. (696) SWIFT *Apus apus* M Bulbul-: 17 cm. Cliffs. The mountain ranges of Pakistan from N Baluchistan to Chitral and east in the Himalayas through Kashmir and Ladakh (up to 5700 m) to Lahul; *c.* 1500-3300 m, foraging to over 4000 m. Also recorded from Assam, Gujarat and the Andaman and Maldive Is. (on migration).

7. (699) LARGE WHITERUMPED SWIFT *Apus pacificus* RM Bulbul-: 18 cm. Himalayas; Assam; NE hill states; SW India (Bombay, Konkan, N Kanara and Malabar), also Gujarat and Andhra Pradesh.

8. (685) INDIAN EDIBLE-NEST SWIFTLET *Collocalia unicolor* R Sparrow-: 12 cm. Caves in rocky off-shore islets and cliffs. SW India from S Konkan south through Karnataka, Kerala and Tamil Nadu; Sri Lanka; up to *c.* 2200 m.

9. (698) DARKBACKED SWIFT *Apus acuticauda* RM Bulbul-: 17 cm. Vicinity of high cliffs. Meghalaya, and presumably Mizoram. Nepal (type specimen in 1865).

10. (691) LARGE BROWNTHROATED SPINETAIL SWIFT *Chaetura gigantea* R Myna±: 23 cm. Evergreen and moist-deciduous biotope. NE hill states; W Ghats complex from Goa south to Kerala, and east to Bangalore and Salem (Tamil Nadu); also Sri Lanka and Andaman Is.

11. (690) COCHINCHINA SPINETAIL SWIFT *Chaetura cochinchinensis* R Bulbul±: 20 cm. C Nepal terai (*c.* 600 m); NE hill states; Bangladesh (Chittagong hill tracts).

12. (693) ALPINE SWIFT *Apus melba* RM Bulbul+: 22 cm. High cliffs. N Baluchistan; Himalayas; W Ghats from Nasik to Kerala; Madhya Pradesh; Sri Lanka.

13. (707) PALM SWIFT *Cypsiurus parvus* R Sparrow-: 13 cm. Associated with palms, especially *Borassus*. India from Uttar Pradesh and Gujarat east through Bhutan, NE India and Bangladesh to Burma; from the Nepal duns south through the Peninsula; Sri Lanka.

14. (688) WHITETHROATED SPINETAIL SWIFT *Chaetura caudacuta* R Bulbul±: 20 cm. Vicinity of cliffs. Himalayas; NE hill states to Bangladesh; *c.* 1200-4000 m.

PLATE 57

1. (715) REDHEADED TROGON *Harpactes erythrocephalus* R Myna+ with longer tail: overall 35 cm. Evergreen forest. Himalayas (foothills up to *c.* 2400 m) from Kumaon to Bhutan and Arunachal Pradesh; NE hill states; Bangladesh eastwards to Burma.

2. (716) WARD'S TROGON *Harpactes wardi* R Pigeon± with longer tail: overall 40 cm. Tall subtropical forest. E Himalayas, between *c.* 1500 and 3000 m.

3. (712) MALABAR TROGON *Harpactes fasciatus* R Myna± with longer tail: overall 31 cm. Moist-deciduous and evergreen forest. From S Gujarat and Khandesh east to Madhya Pradesh and Orissa, and south through the Peninsula in E and W Ghat complexes; Sri Lanka.

4. (744) CHESTNUTHEADED BEE-EATER *Merops leschenaulti* R Bulbul±: 21 cm. Mixed moist-deciduous forest near streams. Himalayan foothills from Kumaon to Bhutan and Arunachal Pradesh; NE hill states; Bangladesh; E & W Ghats; south to Kerala; Sri Lanka; Andaman and Coco Is.

5. (753) BLUEBEARDED BEE-EATER *Nyctyornis athertoni* R Pigeon+: 36 cm. Secondary evergreen and moist-deciduous forest. Himalayan foothills from Himachal Pradesh east to Arunachal Pradesh; Assam; NE hill states; Bangladesh. Also E and W Ghat complexes, and suitable biotopes in peninsular India.

6. (747) BLUECHEEKED BEE-EATER *Merops superciliosus* RM Bulbul+: overall 31 cm. Near jheels and tanks, and sandy seashores. Pakistan south to Sind, and NW India east to Kutch, Rajasthan and Delhi.

7. (746) EUROPEAN BEE-EATER *Merops apiaster* RM Myna+: 27 cm. Open country, cultivation and near lakes, etc. *Breeding:* Baluchistan, north to Chitral, east to Kashmir; between *c.* 900 and 2100 m, foraging up to 3600 m. Passage migrant through the Makran coast and W Himalayas, Garhwal, Sind, Punjab and Rajasthan. Straggler in Karnataka, Tamil Nadu, Maldives.

8. (754) EUROPEAN ROLLER *Coracias garrulus* RM Pigeon±: 31 cm. Wooded parkland and cultivation. *Breeding:* From Baluchistan to Chitral and Kashmir. On outbound passage migration through NW and W & SW India and Deccan in autumn.

9. (748) BLUETAILED BEE-EATER *Merops philippinus* RM Bulbul+: overall 31 cm. Open country near streams, jheels, and reservoirs. Breeds and/or seasonal in N Pakistan, N and C India, and patchily over the Subcontinent; Sri Lanka; Andaman and Nicobar Is.

10. (750) GREEN BEE-EATER *Merops orientalis* R Sparrow+: 21 cm. Open country and cultivation. Subcontinent; Bangladesh; Sri Lanka.

11. (755) INDIAN ROLLER *Coracias benghalensis* R Pigeon±: 31 cm. Open country, cultivation, gardens, light deciduous forest, etc. Subcontinent, Lakshadweep; Bangladesh; Sri Lanka.

12. (759) BROADBILLED ROLLER *Eurystomus orientalis* R Pigeon±: 31 cm. Cultivation clearings in evergreen and semi-evergreen forest. SW India from W Karnataka south through Kerala and W Tamil Nadu; Himalayan foothills from Haryana east to Bhutan and Arunachal Pradesh; NE hill states; Bangladesh; Sri Lanka (rare); Andaman Is.

1 ♀
♂
2 ♀
♂
3 ♀
♂
7 ↓
5
4
7
6
6 ↓
9
10
8 ↑
8
12 ↓
11 ↓
11
12
John H. Dick

1
2
3
3 ↓
4
5
6
6 ↓
7
8
9
10
11
12
13
13 ↑
14
14
14 ↓
John H. Dick

PLATE 58

1. (717) HIMALAYAN PIED KINGFISHER *Ceryle lugubris* R Crow±: 41 cm. Along streams in well-wooded areas. Lower Himalayas from Kashmir to Bhutan and Arunachal Pradesh; NE hill states; Bangladesh; up to *c.* 2000 m.

2. (727a) THREETOED KINGFISHER *Ceyx erithacus rufidorsus*? Sparrow-: 13 cm. Sikkim (once). May be an aberrant polymorphic example of *erithacus*.

3. (719) LESSER PIED KINGFISHER *Ceryle rudis* R Myna+: 31 cm. Stagnant water, jheels, resrvoirs, ditches and slow-flowing streams. Subcontinent; Sri Lanka.

4. (727) THREETOED KINGFISHER *Ceyx erithacus* R Sparrow-: 13 cm. Shady streams in moist-deciduous and evergreen biotope. Nepal and E India; W Ghats complex from about Belgaum south; Andaman and Nicobar Is; Sri Lanka.

5. (725) BLUE-EARED KINGFISHER *Alcedo meninting* R Sparrow±: 16 cm. Small streams in evergreen or bamboo forest. E Himalayan foothills; E India (Bihar, Orissa); W Ghats complex south of Belgaum; Andaman Is; Sri Lanka.

6. (722) COMMON or SMALL BLUE KINGFISHER *Alcedo atthis* RM Sparrow+: 18 cm. Along streams, canals, ponds, mangrove swamps and seashore. Subcontinent; Sri Lanka.

7. (721) BLYTH'S or GREAT BLUE KINGFISHER *Alcedo hercules* R Myna-: 20 cm. Shady streams in dense evergreen jungle. E Himalayan foothills (up to *c.* 1200 m); NE hill states to Bangladesh.

8. (739) BLACKCAPPED KINGFISHER *Halcyon pileata* R Myna+: 30 cm. Along seashore, mangrove swamps, creeks and estuaries. Coastal Bangladesh and round the peninsular seaboard west to Bombay; Andaman and Nicobar Is.; Sri Lanka. Once in Madhya Pradesh (Kanha, March).

9. (729) BROWNWINGED KINGFISHER *Pelargopsis amauroptera* R Pigeon+: 36 cm. Tidal rivers, creeks and mangrove swamps. Coastal Bangladesh, W Bengal and Orissa.

10. (740) WHITECOLLARED KINGFISHER *Halcyon chloris* R Myna±: 24 cm. Mangrove swamps. Coastal Maharashtra and possibly further south; coast of Orissa, W Bengal and Bangladesh; Andaman and Nicobar Is.

11. (733) RUDDY KINGFISHER *Halcyon coromanda* R Myna+: 26 cm. Dense evergreen jungle at streams and pools, and mangrove swamps. The terai and duars from Nepal to Assam; NE hill states to Bangladesh and W Bengal; up to *c.* 1800 m; Andaman Is.

12. (730) STORKBILLED KINGFISHER *Pelargopsis capensis* R Pigeon+: 38 cm. Well-wooded country near streams and irrigation channels. N and peninsular India east of Dehra Dun and south of Himalayan foothills to Kerala; Sri Lanka; Andaman and Nicobar Is.

13. (735) WHITEBREASTED KINGFISHER *Halcyon smyrnensis* R Myna+: 28 cm. Canals, streams, reservoirs, cultivation, gardens and edges of forest often away from water. Subcontinent, Andamans, Nicobars; Sri Lanka.

14. (763) HOOPOE *Upupa epops* RM Myna+: 31 cm. Open hillsides, cultivation, light forest and near villages. Subcontinent, Andamans; Sri Lanka.

PLATE 59
1 ♂
2 ↓
2
1 ♀
4
5
3
4 ↓
8 ♂
8 ♀
7 ♀
1 ♀ ↓
6
8 ↓
3 ↓
7 ♂ ↓
6 ↓
7 ♂
9 ♀
9 ↓
9 ♂
John M. Dick

PLATE 59

1. (773) NARCONDAM HORNBILL *Rhyticeros plicatus* R (endemic) Kite+: Male 66 cm; Female smaller. Narcondam I., Andamans.

2. (767) COMMON GREY HORNBILL *Tockus birostris* R Kite±: 61 cm. Deciduous forest and groves near cultivation. Pakistan (Punjab) and India from Himalayan foothills south to Kerala.

3. (770) WHITETHROATED BROWN HORNBILL *Ptilolaemus tickelli* R Kite+: Male 76 cm; Female smaller. Evergreen forest. E. Assam, Arunachal Pradesh(?), Nagaland(?), Manipur(?). Up to *c.* 900 m.

4. (776) GREAT PIED HORNBILL *Buceros bicornis* R Vulture+: Male 130 cm; Female smaller. Evergreen and moist-deciduous forest. Lower Himalayas (up to c. 2100 m.) from Kumaon east to Bhutan and Arunachal Pradesh; NE hill states; Bangladesh (Chittagong hill tracts); W Ghats complex from Khandala south through Kerala and W. Tamil Nadu, up to *c.* 1500 m.

5. (768) MALABAR GREY HORNBILL *Tockus griseus* R Kite±: 59 cm. Open evergreen and moist-deciduous forest. Western Ghats complex from Bombay south to Kerala and W Tamil Nadu; Sri Lanka, up to *c.* 1600 m.

6. (774) INDIAN PIED HORNBILL *Anthracoceros malabaricus* R Vulture-: Male 89 cm; Female smaller. Moist-deciduous and remnant evergreen forest. From Haryana through Uttar Pradesh, Nepal, Sikkim, Bengal, Bhutan, Arunachal Pradesh, south to Manipur and the hills of Bangladesh and Bihar; E Ghats south to Bastar district and E Andhra Pradesh.

7. (772) WREATHED HORNBILL *Rhyticeros undulatus* R Vulture+: Male 114 cm; Female, with shorter bill, 98 cm. Evergreen forest and edges of clearings up to *c.* 2400 m. Sub-Himalayan duars and foothills from N Bengal and Bhutan east to Arunachal Pradesh; NE hill states; Bangladesh (Chittagong hill tracts).

8. (775) MALABAR PIED HORNBILL *Anthracoceros coronatus* R Vulture±: Male 92 cm; Female smaller. Remnant evergreen and moist-deciduous forest. SE India from Uttar Pradesh and Bihar south through Orissa and Andhra Pradesh; also in the SW from Ratnagiri south through Karnataka, Tamil Nadu and Kerala; Sri Lanka.

9. (771) RUFOUSNECKED HORNBILL *Aceros nipalensis* R Vulture+: 122 cm. Tall evergreen forest up to *c.* 1800 m. E Himalayan foothills; NE hill states; Bangladesh (Chittagong hill tracts).

PLATE 60

1. (784) LINEATED BARBET *Megalaima lineata* R Myna+: 28 cm. Moist-deciduous and light secondary forest, and gardens. Lower Himalayas (up to *c.* 1000 m) from Dehra Dun east through Arunachal Pradesh; NE hill states; Bangladesh; N Orissa.

2. (785) SMALL GREEN BARBET *Megalaima viridis* R Myna±: 23 cm. Evergreen and moist-deciduous forest, gardens and groves. W Ghats complex from Narmada river south through Kerala; Shevaroy and Chitteri hills.

3. (782) LARGE GREEN BARBET *Megalaima zeylanica* R Myna+: 27 cm. Well-wooded moist- and dry-deciduous country, often entering gardens within city limits. Subcontinent; Sri Lanka.

4. (778) GREAT HILL BARBET *Megalaima virens* R Myna+: 33 cm. Subtropical evergreen and moist-temperate forest, *c.* 1000-3000 m. Himalayas; NE hill states; Bangladesh (Chittagong region).

5. (788) BLUETHROATED BARBET *Megalaima asiatica* R Myna±: 23 cm. Light deciduous and evergreen forest and groves. Lower Himalayas (up to *c.* 2000 m) from Rawalpindi and Kashmir east to Arunachal Pradesh; south to Manipur, Chittagong and Calcutta.

6. (787) GOLDENTHROATED BARBET *Megalaima franklinii* R Myna±: 23 cm. Evergreen forest between *c.* 600 and 2400 m. E Himalayas; NE hill states; Bangladesh (Chittagong region).

7. (786) YELLOWFRONTED BARBET *Megalaima flavifrons* R Myna-: 21 cm. Light forest and orchards. Sri Lanka.

8. (792) CRIMSONBREASTED BARBET or COPPERSMITH *Megalaima haemacephala* R Sparrow+: 17 cm. Dry- and moist-deciduous biotope. Subcontinent; Sri Lanka.

9. (790) CRIMSONTHROATED BARBET *Megalaima rubricapilla malabarica* R Sparrow+: 17 cm. Evergreen forest up to *c.* 1200 m. W Ghats complex from Goa south through Kerala and W Tamil Nadu.

10. (791) CRIMSONTHROATED BARBET *Megalaima rubricapilla rubricapilla* R Sparrow+: 17 cm. Open wooded country. Sri Lanka.

11. (789) BLUE-EARED BARBET *Megalaima australis* R Sparrow±: 17 cm. Dense evergreen forest up to *c.* 1200 m. E Himalayan foothills, terai and duars; NE hill states; Bangladesh (Chittagong region).

12. (858) BLACKBACKED WOODPECKER *Chrysocolaptes festivus* R Pigeon-: 29 cm. Deciduous forest. From Dehra Dun, C Rajasthan and E Gujarat to lower Bengal, and from Oudh terai, SW Nepal and S Bihar, Madhya Pradesh, Maharashtra to Kerala; Sri Lanka.

13. (819) LESSER GOLDENBACKED WOODPECKER *Dinopium benghalense* R Myna+: 29 cm. Dry- and moist-deciduous biotope. Pakistan and the rest of the Subcontinent; Bangladesh; Sri Lanka.

14. (823) Sri Lanka ssp. *psarodes* of 819.

15. (824) HIMALAYAN GOLDENBACKED THREETOED WOODPECKER *Dinopium shorii* R Pigeon±: 31 cm. Tall deciduous and semi-evergreen forest up to *c.* 2900 m. Himalayan foothills and terai from Haryana eastward; NE hill states; also peninsular hills (very locally) and hills of Bangladesh.

16. (825) INDIAN GOLDENBACKED THREETOED WOODPECKER *Dinopium javanense* R Myna±: 28 cm. Moist-deciduous and evergreen forest. SW India from Goa south through Kerala and adjacent hills of Tamil Nadu; Assam, NE hill states (?), Bangladesh (Chittagong region).

17. (861) LARGER GOLDENBACKED WOODPECKER *Chrysocolaptes lucidus* R Pigeon±: 33 cm. Evergreen, semi-evergreen and moist-deciduous forest. Himalayas: Garhwal through Nepal east to Bhutan and Arunachal Pradesh; NE hill states; Bangladesh; Bengal, Bihar and E Uttar Pradesh to Orissa; N Andhra Pradesh and SE Madhya Pradesh; W Ghats complex south from the Tapti river to Kerala and W Tamil Nadu.

18. (863) Sri Lanka ssp. *stricklandi* of 861.

1
2
3
4
5
6
7
8
9
10
11
♀
12 ♂
13 ♂
14 ♂
15 ♂
16 ♂
17 ♂
18 ♂
John H. Dick

PLATE 61
1
♀
2
3
4 ♂
5
6 ♂
7 ♂
8 ♂
9 ♂
10 ♂
11 ♂
12 ♂
13 ♂
16 ♂
14 ♂
15 ♂
18 ♂
17 ♂
John H. Dick

PLATE 61

1. (851) PIGMY WOODPECKER *Picoides nanus* R Sparrow-: 14 cm. Groves, deciduous and semi-evergreen forest, up to *c.* 1200 m. Subcontinent (but absent in Pakistan); Bangladesh; Sri Lanka.

2. (798) SPECKLED PICULET *Picumnus innominatus* R Sparrow-: 10 cm. Moist-deciduous and semi-evergreen forest up to 2000 m. Lower Himalayas from Hazara (Pakistan) and Kashmir east to Arunachal Pradesh; Assam; NE hill states; Bangladesh; W and E Ghat complexes; Orissa; lower Bengal.

3. (850) GREYCROWNED PIGMY WOODPECKER *Picoides canicapillus* R Sparrow-: 14 cm. Open tropical semi-evergreen forest up to *c.* 1700 m. Lower Himalayas; NE hill states; Bangladesh.

4. (856) HEARTSPOTTED WOODPECKER *Hemicircus canente* R Sparrow+: 16 cm. Moist-deciduous and secondary evergreen forest. W Ghats complex from Kerala north to Tapti river and east to Madhya Pradesh and Orissa; Bangladesh; NE hill states.

5. (800) RUFOUS PICULET *Sasia ochracea* R Sparrow-: 9 cm. Semi-evergreen and deciduous secondary scrub and bamboo jungle. Lower Himalayas, Garhwal east to Arunachal Pradesh; NE hill states; Bangladesh; up to *c.* 2100 m.

6. (842) BROWNFRONTED PIED WOODPECKER *Picoiaes auriceps* R Bulbul±: 20 cm. Pine, oak and deodar forest. Baluchistan north to Chitral and east along the Himalayas to Nepal; *c.* 1000-3100 m.

7. (855) THREETOED WOODPECKER *Picoides tridactylus* R Myna±: 23 cm. Conifer and deciduous forest. Arunachal Pradesh, between *c.* 3000 and 3800 m.

8. (844) STRIPEBREASTED PIED WOODPECKER *Picoides atratus* R Bulbul±: 21 cm. Pine and oak forest up to *c.* 1500 m. Assam and NE hill states.

9. (847) YELLOWFRONTED PIED WOODPECKER *Picoides mahrattensis* R Bulbul-: 18 cm. Semi-desert to moist-deciduous biotope. Subcontinent; west to Indus river. Sri Lanka.

10. (845) FULVOUSBREASTED PIED WOODPECKER *Picoides macei* R Bulbul-: 19 cm. All types of open forest and wooded country up to *c.* 2000 m. Himalayan foothills and terai east to Bhutan; E India; Bangladesh, Bengal, Bihar and Orissa to Andhra Pradesh; Andaman Is.

11. (833) RUFOUSBELLIED WOODPECKER or SAPSUCKER *Hypopicus hyperythrus* R Bulbul±: 20 cm. Subtropical pine and moist temperate forest. Hazara, Pakistan; Kashmir and Ladakh east through Kumaon to Arunachal Pradesh; NE hill states; Bangladesh; *c.* 800-4100 m.

12. (835) SIND PIED WOODPECKER *Picoides assimilis* R Myna-: 22 cm. Dry scrub jungle up to *c.* 1000 m. Indus valley; Baluchistan (up to *c.* 2000 m.); Salt Range and foothills of Kohat and Rawalpindi.

13. (840) CRIMSONBREASTED PIED WOODPECKER *Picoides cathpharius* R Bulbul-: 18 cm. Moist-deciduous and evergreen forest. E Himalayas; NE hill states. Optimum zone 1800-3000 m.

14. (834) GREAT SPOTTED or REDCROWNED PIED WOODPECKER *Picoides major* R Myna±: 24 cm. Oak, pine and subtropical wet forest. Assam; NE hill states; *c.* 2000-3000 m.

15. (837) HIMALAYAN PIED WOODPECKER *Picoides himalayensis* R Myna±: 25 cm. Hill forest of fir, oak, etc. Kashmir, Himachal Pradesh, Garhwal, Kumaon and W Nepal; *c.* 1500 m and above.

16. (836) Ssp. *albescens* of 837. Safed Koh, Chitral, Gilgit to Himachal Pradesh.

17. (EL) WHITEMANTLED WOODPECKER *Dendrocopos leucopterus* Central Asia.

18. (838) DARJEELING PIED WOODPECKER *Picoides darjellensis* R Myna±; 25 cm. Pine, oak, rhododendron and subtropical wet forest, from *c.* 1700 m to timber-line. E. Himalayas; NE hill states.

PLATE 62

1. (796) WRYNECK *Jynx torquilla* M Bulbul-: 19 cm. Thorn jungle, open deciduous scrub and cultivation. Subcontinent; Sri Lanka (absent).

2. (815) SMALL YELLOWNAPED WOODPECKER *Picus chlorolophus* R Myna+: 27 cm. Mixed deciduous and evergreen secondary jungle. Himalayas up to *c.* 2100 m; E India; Bangladesh; Bihar.

3. (816) Ssp. *chlorigaster* of 815. Peninsular India south of the Satpuras, Orissa; Sri Lanka.

4. (813) LARGE YELLOWNAPED WOODPECKER *Picus flavinucha* R Pigeon±: 33 cm. Mixed evergreen and deciduous forest. Himalayan foothills up to *c.* 2400 m; E India, Bangladesh and hills of S Bihar and Orissa.

5. (809) BLACKNAPED GREEN WOODPECKER *Picus canus* R Pigeon±: 32 cm. Moist subtropical and temperate forest. Himalayas, 2100-3500 m; NE hill states; Bangladesh to Orissa.

6. (807) SCALYBELLIED GREEN WOODPECKER *Picus squamatus* R Pigeon+: 35 cm. Open evergreen mixed oak and pine forest, and orchards. N Baluchistan; W Himalayas: *c.* 1000-3300 m.

7. (808) LITTLE SCALYBELLIED GREEN WOODPECKER *Picus myrmecophoneus* R Myna+: 29 cm. Semi-evergreen, moist-deciduous and sal forest. Haryana, Rajasthan and Gujarat east to Bangladesh, south to Karnataka (Bangalore); Sri Lanka.

8. (830) INDIAN GREAT BLACK WOODPECKER *Dryocopus javensis* R Crow+: 48 cm. Evergreen and moist-deciduous forest, up to *c.* 1200 m. The entire W Ghats complex south to Kerala; E Madhya Pradesh; E Ghats complex(?).

9. (831) Andaman ssp. of 830. *Dryocopus javensis hodgei* Restricted to Andaman Is.

10. (828) HIMALAYAN GREAT SLATY WOODPECKER *Mulleripicus pulverulentus* R Crow+: 51 cm. Sal and tropical semi-evergreen forest. Himalayan terai and foothills up to 1200 m; NE hill states; Bangladesh.

11. (EL) LACED WOODPECKER *Picus vittatus* SE Asia.

12. (827) PALEHEADED WOODPECKER *Gecinulus grantia* R Myna±: 25 cm. Moist-deciduous bamboo and scrub jungle up to *c.* 1000 m. E Himalayas; Assam; NE hill states; Bangladesh (hills).

13. (857) REDEARED BAY WOODPECKER *Blythipicus pyrrhotis* R Pigeon-: 27 cm. Dense forest. E Himalayas up to *c.* 2000 m; NE hill states; Bangladesh.

14. (804) RUFOUS WOODPECKER *Micropternus brachyurus* R Myna+: 25 cm. Moist-deciduous biotope. Subcontinent; Bangladesh; Sri Lanka.

1
2 ♂
3 ♂
4 ♂
5 ♂
6 ♂
7 ♂
8 ♂
9 ♂
10 ♂
11 ♂
12 ♂
13 ♂
14 ♂
John H. Dick

PLATE 63
1
2 ♂
♀
imm
3
imm
4
5 ♂
♀
6 ♂
7 ♂
♀
8 ↓
9 ↓
8 ♂
9
10 ♂
11 ↓
11
12 ↓
12

PLATE 63

1. (865) LONGTAILED BROADBILL *Psarisomus dalhousiae* R Bulbul+: 27 cm. Tropical and subtropical evergreen forest. E Himalayas west to Garhwal; NE hill states; Bangladesh (hills); up to *c.* 2000 m.

2. (952) GOLDEN ORIOLE *Oriolus oriolus* RM Myna+: 25 cm. Deciduous and semi-evergreen biotope: groves, orchards, cultivation, gardens, etc. Subcontinent.

3. (864) COLLARED BROADBILL *Serilophus lunatus* R Bulbul±: 19 cm. Tropical evergreen and semi-evergreen forest. E Himalayas; NE hill states; Bangladesh (hills); up to *c.* 1700 m.

4. (958) BLACKHEADED ORIOLE *Oriolus xanthornus* R Myna+: 25 cm. Tropical moist-deciduous biotope: Light forest, plantations, village groves, gardens, etc. Subcontinent; Sri Lanka; Andaman Is.

5. (954) BLACKNAPED ORIOLE *Oriolus chinensis* RM Myna+: 25 cm. Mixed deciduous and evergreen jungle, plantations, gardens. Subcontinent; Andaman and Nicobar Is.; Sri Lanka (a 100 year old record).

6. (955) SLENDERBILLED BLACKNAPED ORIOLE *Oriolus chinensis tenuirostris* RM Myna+: 25 cm. Pine forest and semi-evergreen jungle. *Breeding*: E Himalayan foothills west to Bhutan. *Winter*: Bangladesh; Bihar; C Nepal; *c.* 1500-2100 m.

7. (961) MAROON ORIOLE *Oriolus traillii* R Myna+: 28 cm. Moist-deciduous and evergreen forest. Lower Himalayas; NE hill states; Bangladesh (Chittagong region); up to *c.* 2400 m.

8. (871) BLUE PITTA *Pitta cyanea* R Quail+: 23 cm. Evergreen and bamboo forest. E Himalayas; NE hill states; Bangladesh (hills); up to *c.* 2000 cm.

9. (869). HOODED or GREENBREASTED PITTA *Pitta sordida* R Quail±: 19 cm. Moist-deciduous and evergreen secondary forest and scrub jungle. From C Nepal and lower Bengal east; up to *c.* 2000 m; Nicobar Is. Obtained at Simla; Dehra Dun (sight).

10. (866) BLUENAPED PITTA *Pitta nipalensis* R Quail+: 25 cm. Tropical and subtropical secondary evergreen and bamboo jungle. E Himalayas; NE hill states; Bangladesh (hills); up to *c.* 2000 m.

11. (868) BLUEWINGED PITTA *Pitta moluccensis* V Quail+: 23 cm. Single record, Bangladesh (Barisal; also Sunderbans).

12. (867) INDIAN PITTA *Pitta brachyura* R Quail±; 19 cm. Scrub jungle, light deciduous or dense evergreen forest. Subcontinent; Sri Lanka.

PLATE 64

1. (872) SINGING BUSH LARK *Mirafra javanica* R Sparrow±: 15 cm. Grassland, fallow cultivation and sparse scrubby semi-desert. Pakistan and most of the Subcontinent.

2. (873) BUSH LARK *Mirafra assamica* R Sparrow±: 15 cm. Semi-desert, sparse scrub jungle, grassland and cultivation. From Haryana eastwards, Arunachal, south to N Orissa and N Madhya Pradesh.

3. (874) Ssp. *affinis* of 873. S Orissa, SE Madhya Pradesh and Belgaum south; Sri Lanka.

4. (877) REDWINGED BUSH LARK *Mirafra erythroptera* R Sparrow±. 14 cm. Sparse scrub jungle and fallow land at low elevations. Plains of Pakistan, NW India and most of the Subcontinent.

5. (878) ASHYCROWNED FINCH-LARK *Eremopterix grisea* R Sparrow-: 13 cm. Sparsely scrubbed waste land, stubble and ploughed fields. Subcontinent; Sri Lanka.

6. (879) BLACKCROWNED FINCH-LARK *Eremopterix nigriceps* R Sparrow-; 13 cm. Sandy waste land near cultivation. Plains of Pakistan and NW India to Delhi, Rajasthan and Kathiawar.

7. (880) DESERT FINCH-LARK *Ammomanes deserti* R Sparrow±: 16 cm. Fallow land and barren stony ground. Pakistan east to Sind, Bahawalpur, Punjab and India in Jammu and Kashmir; up to *c.* 2000 m.

8. (884) BIFASCIATED or LARGE DESERT LARK *Alaemon alaudipes* R Myna±: 23 cm. Sandy desert and saltflats. Pakistan; Baluchistan to Bahawalpur and Sind. India: Rann of Kutch.

9. (882) RUFOUSTAILED FINCH-LARK *Ammomanes phoenicurus* R Sparrow±: 16 cm. Open stony scrub-and-bush plains and plateau country, fallow land and near cultivation. Pakistan, N India and most of the Subcontinent except the NE, and Kerala. Absent also in Sri Lanka.

10. (881) BARTAILED DESERT LARK *Ammomanes cincturus* R Sparrow±: 16 cm. Barren rocky ground. Baluchistan and W Sind; up to *c.* 1500 m.

11. (892) EASTERN CALANDRA LARK *Melanocorypha bimaculata* M Sparrow+: 16 cm. Semi-desert, fallow fields, mudflats and margins of jheels. Pakistan and NW India.

12. (EL) CALANDRA LARK *Melanocorypha calandra* Mediterranean region and W Asia.

13. (888a) LESSER SHORT-TOED LARK *Calandrella rufescens* M Sparrow±: 15 cm. Semi-desert and waste land. Winter visitor to Pakistan. Passage migrant in Gilgit.

14. (886) SHORT-TOED LARK *Calandrella cinerea* M Sparrow±: 15 cm. Stony grassland, sandy semi-desert and saltflats. Subcontinent. Absent in Sri Lanka, Andaman and Nicobar Is.

15. (891) SAND LARK *Calandrella raytal* R Sparrow-: 13 cm. Sandy river banks and islets. From Peshawar and the Makran coast to Kutch; throughout the entire Ganges and Brahmaputra river systems in N India, Assam, Bangladesh and Bengal; Madhya Pradesh south to Hoshangabad and Mhow.

16. (894) LONGBILLED CALANDRA LARK *Melanocorypha maxima* R Myna-: 21 cm. Marshes and bogs around lakes. Sikkim and Bhutan above *c.* 3600 m; Rupshu (Ladakh) at *c.* 4300 cm.

17. (897) HORNED LARK *Eremophila alpestris* R Bulbul±: 20 cm. Barren stony ground in Tibetan steppe facies. Hazara and Karakoram east to Bhutan and Arunachal Pradesh. Seasonally, *c.* 3000-5500 m.

18. (895) Ssp. *albigula* of 897. Chitral and Gilgit.

19. (902) SYKES'S CRESTED LARK *Galerida deva* R Sparrow±: 13 cm. Stony, sparsely scrubbed plateau country and dry cultivation. S peninsular India.

20. (899) CRESTED LARK *Galerida cristata* R Sparrow+: 18 cm. Sandy semi-desert and cultivated plains. Pakistan; N India and NE hill states.

21. (901) MALABAR CRESTED LARK *Galerida malabarica* R Sparrow±: 15 cm. Sparse scrub jungle, grassy stony ground and semi-cultivation. W. Peninsular India from Gujarat south to Kerala and Tamil Nadu.

22. (907) EASTERN SKYLARK *Alauda gulgula* R Sparrow±: 16 cm. Grassland and young crops. Subcontinent; Sri Lanka. Absent in Andaman and Nicobar Is.

23. (903) SKYLARK *Alauda arvensis* M Sparrow+: 18 cm. Grassland and cultivation. Pakistan and N India to Uttar Pradesh and C Nepal.

6 u.w.
5 underwing
1
2
3
4
5 ♀
5 ♂
6 ♂
7
8
9
10
11
12
13
14
15
16
17 ♂
18 ♂
19
20↑
20
21
22
23

PLATE 65
1
2
3
4
5
6
7
8
9
10
11
12
13
14
15
16
17
18
John H. Dick

PLATE 65

1. (709) CRESTED TREE SWIFT *Hemiprocne longipennis* R Bulbul±: 23 cm. Well-wooded country. Subcontinent excepting Pakistan; Sri Lanka.

2. (910) COLLARED SAND MARTIN *Riparia riparia* RM Sparrow-: 13 cm. Streams and lakes. *Breeding*: Himalayas from Quetta eastwards; NE hill states; W Bengal; up to 4500 m. *Winter*: Rest of Pakistan, south to Makran; Madhya Pradesh, Bihar; Sri Lanka.

3. (912) PLAIN SAND MARTIN *Riparia paludicola* R Sparrow-: 12 cm. River valleys. Pakistan and N Subcontinent roughtly south to Bombay, Madhya Pradesh, Orissa to NE hill states; Bangladesh.

4. (915) PALE CRAG MARTIN *Hirundo obsoleta* R Sparrow-: 13 cm. Vicinity of cliffs in bare foothills. Baluchistan; Sind; NWFP; Kutch (winter).

5. (913) CRAG MARTIN *Hirundo rupestris* RM Sparrow±: 14 cm. Around crags and old hill forts. *Breeding*: NW Pakistan hills and through Ladakh and Kashmir to Garhwal, *c.* 1200-4500 m. *Winter*: Pakistan, N, C and W India.

6. (914) DUSKY CRAG MARTIN *Hirundo concolor* R Sparrow-: 13 cm. Vicinity of crags, forts, old buildings, etc. Practically all Subcontinent. Not in Andamans, Nicobars, Bangladesh and Sri Lanka.

7. (932) NEPAL HOUSE MARTIN *Delichon nipalensis* R Sparrow-: 13 cm. River valleys, mountain ridges, etc. Himalayas from Garhwal to Arunachal Pradesh; NE hill states; seasonally *c.* 300-4000 m.

8. (EL) ASIAN HOUSE MARTIN *Delichon dasypus* S and E Asia: Indonesia, etc.

9. (921) WIRETAILED SWALLOW *Hirundo smithii* R Sparrow±: 14 cm. Open country, cultivation, near water. Subcontinent. Not in Andamans and Nicobars. Sri Lanka (once).

10. (930) HOUSE MARTIN *Delichon urbica* RM Sparrow±: 15 cm. Around cliffs. Pakistan and Himalayas east to Arunachal Pradesh. Occasionally wintering as far south as Kerala.

11. (916) SWALLOW *Hirundo rustica* RM Sparrow±, with long forked tail: overall 18 cm. Open cultivated country, and around human settlements. *Breeding*: Pakistan and along the Himalayas up to *c.* 3000 m., Assam (Cachar). *Winter*: Subcontinent, Andamans, Nicobars, Lakshadweep; Sri Lanka; Maldives.

12. (918) Migrant spp. *tytleri* of 916. Bangladesh and Bengal eastwards (winter).

13. (919) HOUSE SWALLOW *Hirundo tahitica* R Sparrow-: 13 cm. Grassy slopes and near settlements. W Ghats complex in SW India; Sri Lanka; Andaman Is.

14. (923) STRIATED or REDRUMPED SWALLOW *Hirundo daurica* RM Sparrow±, with longer tail: overall 20-23 cm. Open, broken or cultivated country. Subcontinent, Andamans (?), Car Nicobar; Sri Lanka.

15. (928) Sri Lanka endemic ssp. *hyperythra* of 923.

16. (983) WHITEBREASTED SWALLOW-SHRIKE *Artamus leucorhynchus* R Bulbul±: 18 cm. Open wooded country, forest clearings and rubber plantations. Andaman Is.

17. (922) INDIAN CLIFF SWALLOW *Hirundo fluvicola* R Sparrow-: 12 cm. Open country and cultivation around rivers and canals. Pakistan; N India upto *c.* 84°E. One record each from Tamil Nadu and Sri Lanka.

18. (982) ASHY SWALLOW-SHRIKE *Artamus fuscus* R Bulbul±: 19 cm. Open dry- and moist-deciduous country, closely associated with tad palms. Subcontinent; Sri Lanka.

PLATE 66

1. (1089) ROSY MINIVET *Pericrocotus roseus* R Bulbul-: 18 cm. Deciduous or evergreen lightly wooded country. Himalayas from NWFP east; NE hill states up to 1800 m.; *Winter*: Pakistan and N Indian Plains; sporadically in the peninsula.

2. (1096) WHITEBELLIED MINIVET *Pericrocotus erythropygius* R Sparrow-, with longer tail: 15 cm. Dry-deciduous forest and thorn scrub. N and C and peninsular India.

3. (1093) SMALL MINIVET *Pericrocotus cinnamomeus* R Sparrow-, with longer tail: 15 cm. Deciduous forest, open scrub, groves of trees, etc. Subcontinent.

4. (1088) YELLOWTHROATED MINIVET *Pericrocotus solaris* R Bulbul-: 17 cm. Open deciduous or evergreen forest. Himalayas from Garhwal eastwards; NE hill states; *c.* 1500-3000 m.

5. (1084) SHORTBILLED MINIVET *Pericrocotus brevirostris* R Bulbul-: 17 cm. Open deciduous forest, edges of evergreen, and secondary growth. E Himalayas; NE hill states; Bangladesh; *c.* 1800-2400 m.

6. (1085) LONGTAILED MINIVET *Pericrocotus ethologus* R Bulbul-: 18 cm. Open deciduous forest. *Breeding*: Lower Himalayas; NE hill states; Bangladesh (Chittagong). *Winter*: Himalayan foothills and NW Subcontinent. Straggler to Sind, Madhya Pradesh, Gujarat and Maharashtra.

7. (1089a) ASHY MINIVET *Pericrocotus divaricatus* V Bulbul-: 18 cm. Five records: Andaman Is., Bombay, Madras City (TN), Kerala and Himachal Pradesh.

8. (1081) SCARLET MINIVET *Pericrocotus flammeus* R Bulbul±: 20 cm. Deciduous, mixed and evergreen forest: Plains and hills. Lower Himalayas, NE hill states, E and SW peninsular India, Andamans; Bangladesh; Sri Lanka.

9. (1080) Ssp. *speciosus* of 1081. Himalayan foothills (to *c.* 2700 m) and adjacent plains; NE hill states; continental India.

10. (1070) COMMON WOOD SHRIKE *Tephrodornis pondicerianus* R Bulbul-: 16 cm. Dry-deciduous scrub, secondary jungle, gardens, etc. Subcontinent; Sri Lanka.

11. (1072) LARGE CUCKOO-SHRIKE *Coracina novaehollandiae* R Pigeon-: 28 cm. Deciduous and mixed forest, scrub and cultivation. Lower Himalayas; NE hill states; most of Subcontinent; Sri Lanka; Andaman Is.

12. (1065) PIED FLYCATCHER-SHRIKE *Hemipus picatus* R Sparrow±: 14 cm. Dry- and moist-deciduous biotope: light forest. Lower Himalayas: Simla to Arunachal Pradesh; NE hill states; Bangladesh; up to *c.* 2250 m; C and peninsular India; Sri Lanka.

13. (1068) LARGE WOOD SHRIKE *Tephrodornis virgatus* R Myna±: 23 cm. Tall trees in evergreen and moist-deciduous forest. W Ghats complex, up to *c.* 1800 m; Himalayas from Uttar Pradesh; NE hill states; Bangladesh up to *c.* 1000 m.

14. (1078) BLACKHEADED CUCKOO-SHRIKE *Coracina melanoptera* RM Bulbul±: 20 cm. Open deciduous forest and scrub jungle. Practically all Subcontinent; Sri Lanka.

15. (1079a) PIED CUCKOO-SHRIKE *Coracina nigra* R Bulbul±: 18 cm. Forest edges and secondary growth. Andaman and Nicobar Is.

16. (1077) SMALLER GREY CUCKOO-SHRIKE *Coracina melaschistos* RM Myna-: *c.* 22 cm. Open forest, pine, oak and chestnut woods and bamboo jungle. *Summer*: Lower Himalayas; NE hill states, 600-2100 m. *Winter*: most of N and peninsular India.

17. (1076) BARRED CUCKOO-SHRIKE *Coracina striata* R Myna±: 26 cm. Heavy moist evergreen forest. Andaman Is.

♀
1 ♂
♀
2 ♂
♀
3 ♂
♀
4 ♂
♀
5 ♂
♀
6 ♂
♀
7 ♂
♀
9 ♂
8 ♂
10
♀
12 ♂
13 ♂
♀
11 ♂
11 ♀
imm
♀
14 ♂
♀
15 ♂
16 ♂
♀
17 ♂
♀
John H. Dick

PLATE 67

PLATE 67

1. (1098) COMMON IORA *Aegithina tiphia* R Sparrow±: 14 cm. Open forest, cultivation and scrub jungle. Subcontinent; Sri Lanka.

2. (1338) FIRETAILED MYZORNIS *Myzornis pyrrhoura* R Sparrow-: 12 cm. Evergreen biotope: rhododendron, juniper and other bushes along streams; heavy jungle and bamboo thickets. E Himalayas *c.* 1600-3950 m.

3. (1102) MARSHALL'S IORA *Aegithina nigrolutea* R Sparrow±: 14 cm. Dry-deciduous scrub, thorn jungle and groves. Pakistan and NW India.

4. (1106) ORANGEBELLIED CHLOROPSIS or LEAF BIRD *Chloropsis hardwickii* R Bulbul±: 19 cm. Open scrub, deciduous and evergreen forest. Himalayas from Simla east to Arunachal Pradesh; NE hill states; Bangladesh; *c.* 600-2600 m.

5. (1103) GOLDENFRONTED CHLOROPSIS or LEAF BIRD *Chloropsis aurifrons* R Bulbul±: 19 cm. Deciduous and evergreen forest and secondary scrub. Himalayan foothills from Garhwal east and Subcontinent (minus Pakistan); Sri Lanka.

6. (1108) GOLDMANTLED CHLOROPSIS or LEAF BIRD *Chloropsis cochinchinensis* R Bulbul-: 18 cm. Deciduous and evergreen forest, gardens, groves, etc. NE hill states; Bangladesh; continental and peninsular India; Sri Lanka.

7. (1120) REDWHISKERED BULBUL *Pycnonotus jocosus* R Myna-: 20 cm. Scrub jungle, light forest and semicultivation. Subcontinent; Andamans; Nicobars (introduced). Absent in Pakistan and Sri Lanka.

8. (1147) BROWNEARED BULBUL *Hypsipetes flavalus* R Myna-: 20 cm. Deep secondary forest, forest edges, cultivation. Himalayan foothills from Kumaon to Arunachal Pradesh; NE hill states; Bangladesh; up to *c.* 1600 m.

9. (1128) REDVENTED BULBUL *Pycnonotus cafer* R Myna-: 20 cm. Cultivation and scrub. Subcontinent; Sri Lanka. Absent in Andaman and Nicobar Is.

10. (1131) Ssp. *bengalensis* of 1128. Kumaon, Uttar Pradesh, Nepal, Sikkim, Bhutan, Arunachal Pradesh, Assam, NE hill states, Bengal, Bihar, Orissa, Bangladesh.

11. (1123) WHITECHEEKED BULBUL *Pycnonotus leucogenys leucotis* R Myna-: 20 cm. Semi-desert biotope: scrub, coastal mangroves and gardens, etc. Pakistan and NW Indian plains.

12. (1125) WHITECHEEKED BULBUL *Pycnonotus leucogenys leucogenys* R Myna-: 20 cm. Open scrub and cultivation, gardens, etc. Himalayan foothills, *c.* 2000-3000 m, from Kashmir east to Arunachal Pradesh.

13. (1148) BLACK BULBUL *Hypsipetes madagascariensis* R Bulbul+: 23 cm. Tall moist-deciduous and evergreen forest. Himalayas; NE hill states; Bangladesh; W Ghats complex south of Bombay; Sri Lanka.

14. (1109) FAIRY BLUEBIRD *Irena puella* R Myna+: 27 cm. Evergreen and moist-deciduous forest. Range disjunct: 1. SW India (W Ghats complex) up to *c.* 1800 m. 2. E Himalayan foothills; NE hill states; hills of Bangladesh; up to *c.* 1200 m. 3. Andaman and Nicobar Is. 4. E Ghats and Orissa. Extinct in Sri Lanka.

PLATE 68

PLATE 68

1. (1111) FINCHBILLED BULBUL *Spizixos canifrons* R Myna-: 20 cm. Deciduous and evergreen forest, scrub and cultivation. NE hill states; Bangladesh (Chittagong hill tracts); seasonally between *c.* 900 and 2500 m.

2. (1133) STRIATED GREEN BULBUL *Pycnonotus striatus* R Myna-: 20 cm. Evergreen jungle, oak and rhododendron forest. E Himalayas; NE hill states; up to *c.* 3000 m.

3. (1114) GREYHEADED BULBUL *Pycnonotus priocephalus* R Bulbul-: 19 cm. Dense evergreen swampy jungle. W Ghats complex south of Goa to Kerala; east to the Nilgiris and Palnis (Tamil Nadu).

4. (1115) BLACKHEADED YELLOW BULBUL *Pycnonotus melanicterus flaviventris* R Bulbul±: 18 cm. Moist-deciduous and evergreen forest and scrub. Himalayan foothills from Kumaon to Arunachal Pradesh; NE hill states; Bangladesh (Chittagong); Orissa, NE Andhra, and E Madhya Pradesh; up to *c.* 1800 m.

5. (1116) Rubythroated ssp. *Pycnonotus melanicterus gularis* R Bulbul±: 18 cm. Evergreen forest and secondary jungle. SW India from Goa south through Kerala and adjacent parts of Tamil Nadu.

6. (1117) Blackcapped ssp. of 1115. *Pycnonotus melanicterus melanicterus* R Bulbul±: 18 cm. Well-wooded areas, open country. Sri Lanka.

7. (1112) BLACKHEADED BULBUL *Pycnonotus atriceps* R Bulbul-: 18 cm. Deciduous forest, light jungle, gardens, etc. NE hill states; Bangladesh; up to *c.* 700 m. Andaman Is.

8. (1135)YELLOWTHROATED BULBUL *Pycnonotus xantholaemus* R Myna-: 20 cm. Sparse thorn scrub jungle interspersed with large trees among broken stony hillocks. The hills of S Andhra Pradesh, E Karnataka and N Tamil Nadu.

9. (1137) BLYTH'S BULBUL *Pycnonotus flavescens* R Myna-: 20 cm. Tree-forest with undergrowth, scrub jungle and secondary growth. NE hill states; Bangladesh (Chittagong region); seasonally between *c.* 450 and 2100 m.

10. (1136) YELLOWEARED BULBUL *Pycnonotus penicillatus* R (endemic) Myna-: 20 cm. Jungle, wooded ravines and well-wooded gardens. Sri Lanka.

11. (1146) RUFOUSBELLIED BULBUL *Hypsipetes mcclellandi* R Myna±: 23 cm. Well-wooded secondary jungle, oak, rhododendron and open pine forest. Outer Himalayas from Kumaon east to Arunachal Pradesh; NE hill states; Bangladesh (Chittagong region); up to *c.* 2700 m.

12. (1141) OLIVE BULBUL *Hypsipetes viridescens* R Myna-: 19 cm. Dense, humid evergreen forest and abandoned cultivation. NE hill states; Bangladesh (Chittagong hill tracts); up to *c.* 900 m.

13. (1138) WHITEBROWED BULBUL *Pycnonotus luteolus* R Myna-: 20 cm. Dry open scrub country and gardens. Midnapur (West Bengal); peninsular India south of Madhya Pradesh; Sri Lanka.

14. (1142) NICOBAR BULBUL *Hypsipetes nicobariensis* R Myna-: 20 cm. Forests and gardens. Nicobar Is.

15. (1144) YELLOWBROWED BULBUL *Hypsipetes indicus* R Myna-: 20 cm. Evergreen jungle, sholas and coffee shade-trees. W Ghats from Pune and Satara district south; Sri Lanka.

16. (1140) WHITETHROATED BULBUL *Cringer flaveolus* R Myna±: 23 cm. Evergreen secondary jungle and shrubby undergrowth in heavy forest. E Himalayas; NE hill states; hills of Bangladesh. Optimum zone 600-1200 m.

PLATE 69

1. (963) BLACK DRONGO or KING-CROW *Dicrurus adsimilis* R Bulbul+: 31 cm. Open deciduous forest and cultivation. Subcontinent; Sri Lanka.

2. (972) LESSER RACKET-TAILED DRONGO *Dicrurus remifer* R Myna+: 28 cm. Heavy moist-deciduous and evergreen forest. Himalayan foothills from Garhwal east to Arunachal Pradesh; NE hill states south to Chittagong (Bangladesh); up to *c.* 2000 m.

3. (970) CROWBILLED DRONGO *Dicrurus annectans* R Myna+: 27 cm. Tropical wet evergreen and semi-evergreen, moist-deciduous and sal forest. Himalayas from Kumaon east to Arunachal Pradesh; Assam and NE hill states; up to *c.* 600 m. Bangladesh.

4. (977) GREATER RACKET-TAILED DRONGO *Dicrurus paradiseus* R Myna±, with longer tail: overall 31-35 cm. Deciduous and evergreen forest, bamboo jungle and edges of cultivation. Subcontinent, Andamans and Nicobars; Sri Lanka; Bangladesh. Absent in Pakistan.

5. (979) Sri Lanka ssp. *lophorhinus* of 977 (outer tail feather).

6. (971) BRONZED DRONGO *Dicrurus aeneus* R Bulbul+: 24 cm. Tropical evergreen and moist-deciduous forest and scrub. Himalayan foothills from Mussoorie east to Arunachal Pradesh; NE hill states; Bangladesh; W Bengal; E and W Ghat complexes; up to *c.* 2000 m.

7. (973) HAIRCRESTED or SPANGLED DRONGO *Dicrurus hottentottus* R Myna±, with longer tail; overall 31 cm. Moist-deciduous and evergreen forest. Himalayan foothills from Himachal Pradesh east to Arunachal Pradesh, south through NE hill states to Bangladesh, Bengal, Bihar, Orissa, and Madhya Pradesh; E & W Ghats complexes north to Bombay; up to *c.* 1400 m. Straggler to Kutch.

8. (965) GREY or ASHY DRONGO *Dicrurus leucophaeus* RM Bulbul+: 30 cm. Moist-deciduous, semi-evergreen and pine forest, plantations, etc. *Breeding*: Himalayas, up to *c.* 3000 m, NE hill states. *Winter*: Well-wooded plains and hills south to Kerala; Andamans (Narcondam I.); Bangladesh; Sri Lanka.

9. (966a) Ssp. *salangensis* of 965. V? Once each from Nagaland and Andaman Is.

10. (975) ANDAMAN DRONGO *Dicrurus andamanensis* R Myna±, with longer tail; overall 35 cm. Evergreen forest. Andaman Is.

11. (588) DRONGO-CUCKOO *Surniculus lugubris* R Myna±. with longer, forked tail; 25 cm. Well-wooded country and orchards. Himalayas; NE hill states; Bangladesh; N and peninsular India; Sri Lanka; up to 1600 m.

12. (967) WHITEBELLIED DRONGO *Dicrurus caerulescens* R Bulbul+: 24 cm. Dry- and moist-deciduous and bamboo forest. Subcontinent; Sri Lanka.

13. (1773) WHITEBREASTED DIPPER *Cinclus cinclus* R Myna-: 20 cm. Swift-flowing streams and torrents and glacial lakes. Himalayas, seasonally between *c.* 2000 and 5100 m.

14. (1775) BROWN DIPPER *Cinclus pallasii* R Myna-: 20 cm. Swift-flowing streams and torrents. Himalayas; NE hill states south to the Chittagong hill tracts (Bangladesh); seasonally between *c.* 450 and 5000 m.

1 2 3 4 5 6 7 8 9 10 11 12 13 14

John H. Dick

PLATE 70

PLATE 70

1. (1022) BLACKTHROATED JAY *Garrulus lanceolatus* R Pigeon±: 33 cm. Deciduous and mixed conifer forest. Himalayas east to Nepal; *c.* 1000-3000 m.

2. (1020) JAY *Garrulus glandarius* R Pigeon±: 33 cm. Wet-temperate mixed forest of oak, pine, etc. Himalayas; NE hill states; seasonally 1000-2500 m. Bangladesh (?).

3. (1029) WHITERUMPED MAGPIE *Pica pica* R Myna±: overall 52 cm (including 30 cm long tail). Cultivation. Pakistan from Chitral south to Baluchistan and west through Ladakh and N Kashmir to Himachal Pradesh, *c.* 1500-4500 m. A black-rumped subspecies (*bottanensis*) in Bhutan and Arunachal Pradesh.

4. (1023) GREEN MAGPIE *Cissa chinensis* R Myna±; with longer tail; overall 47 cm. Tropical and sub-tropical evergreen secondary forest and moist-deciduous bamboo jungle. Himalayas; Assam and NE hill states; Bangladesh (Chittagong); up to *c.* 1600 m.

5. (1024) CEYLON MAGPIE *Cissa ornata* R (endemic) Myna+, with long tail: overall 47 cm. Dense evergreen forest. Sri Lanka, between c. 150 and 2100 m.

6. (1025) YELLOWBILLED BLUE MAGPIE *Cissa flavirostris* R Pigeon±, with long tail (39-43 cm): overall 66 cm. Wet-temperate mixed forest of pine, oak, etc. Himalayas, seasonally between *c.* 1000 and 3300 m.

7. (1027) REDBILLED BLUE MAGPIE *Cissa erythrorhyncha* R Pigeon±, with long tail (*c.* 40 cm) overall 70 cm. Tropical and subtropical dry- and moist-deciduous forest; also around cultivation and villages. Himalayas from Himachal Pradesh to E Nepal (not recorded farther east); Assam, NE hill states; seasonally between 300 and 1800 m.

8. (1038) HIMALAYAN TREE PIE *Dendrocitta formosae* R Myna±, with long tail: overall 41-43 cm. Tropical and subtropical moist-deciduous and semi-evergreen forest and well-wooded country. Himalayas, seasonally between *c.* 600 and 2100 m; NE hill states; Assam and Bengal. An isolated pocket in E Ghats (Vishakhapatnam section).

9. (1036) WHITEBELLIED TREE PIE *Dendrocitta leucogastra* R Myna+, with long tail: overall 50 cm. Wet evergreen biotope. Southern W Ghats complex through Kerala and Tamil Nadu, east to Bangalore and Andhra Pradesh.

10. (1032) INDIAN TREE PIE *Dendrocitta vagabunda* R Myna+, with long tail: overall 50 cm. Dry- and moist-deciduous biotope lightly wooded country, gardens, etc. Subcontinent. Absent in Sri Lanka and Andaman and Nicobar Is.

11. (1035) BLACKBROWED TREE PIE *Dendrocitta frontalis* R Myna±, with long tail: overall 38 cm. Mixed evergreen forest and bamboo jungle. E Himalayas, up to *c.* 2100 m; NE hill states.

12. (1040) ANDAMAN TREE PIE *Dendrocitta bayleyi* R Myna±, with long tail: overall 36 cm. Evergreen forest. Andaman Is.

13. (1042) NUTCRACKER *Nucifraga caryocatactes* R Pigeon±: 32 cm. Subalpine conifer forest, especially pine. From Chitral south and east along Himalayas to Arunachal Pradesh; *c.* 1000 m up to timber-line.

14. (1043) Ssp. *hemispila* of 1042. From Murree east through Kashmir, Nepal and Darjeeling.

15. (1041) HUME'S GROUND CHOUGH *Podoces humilis* R Myna-: 20 cm. Sandy stone-littered hillsides, etc. N Sikkim and N Nepal (Tibetan steppe facies); *c.* 4200-5200 m.

PLATE 71

1. (590) KOEL *Eudynamys scolopacea* R Crow±: 43 cm. Lightly wooded country and cultivation. Subcontinent.

2. (1045) YELLOWBILLED or ALPINE CHOUGH *Pyrrhocorax graculus* R Crow-: 38 cm. High altitude crags, rocky slopes, alpine meadows, etc. From Baluchistan to Chitral; Himalayas; seasonally *c.* 1500-5000 m.

3. (1053) JACKDAW *Corvus monedula* R Pigeon±: 33 cm. Pastures and cultivated fields. *Summer*: Breeding in Kashmir between *c.* 1700 and 2100 m. *Winter*: NWFP, Baluchistan, Gilgit, Punjab south to Ferozepore and east to Haryana.

4. (1046) REDBILLED CHOUGH *Pyrrhocorax pyrrhocorax* R Crow±: 45 cm. Dry- and moist-temperate high mountain biotope. From Baluchistan to Chitral; Himalayas; seasonally *c.* 1200-3500 m - a lower altitudinal range than Yellowbilled.

5. (1052) ROOK *Corvus frugilegus* M Crow+: 48 cm. Cultivated fields. From Chitral and Gilgit south to Baluchistan (Quetta) and east through N Punjab and Haryana.

6. (1049) HOUSE CROW *Corvus splendens* R Pigeon+: 43 cm. Human habitations, towns and cities. Subcontinent. Sri Lanka.

7. (1054) JUNGLE CROW *Corvus macrorhynchos* R Crow+: 48-50 cm. Wooded country and outskirts of towns and cities. Subcontinent, Andamans; Bangladesh; Sri Lanka. Absent in Nicobar Is.

8. (1058) CARRION CROW *Corvus corone* R Crow+: 47 cm. Willow and poplar groves and cultivation with trees and open forest. Baltistan, Ladakh, Kashmir, Murree (*c.* 2400-3600 m), Kohat (*c.* 1500 m to timber-line); Quetta (winter).

9. (1058a) Migrant ssp. *sharpii* of 1058. Cultivated country, *c.* 2500-3500 m. NWFP, Ladakh, Kashmir (winter).

10. (1061) BROWN-NECKED RAVEN *Corvus ruficollis* R Kite±: 58 cm. Flat desert, semi-desert and arid country. Baluchistan and Sind.

11. (1059) RAVEN *Corvus corax* R Kite+: 69 cm. Near towns and villages, desert settlements and semi-desert areas. NWFP, Baluchistan, Sind and desert parts of NW India, and along the higher Himalayas (Tibetan steppe facies) to Arunachal Pradesh. Stragglers recorded in Maharashtra and Madhya Pradesh.

♀
1♂
2
3
3↑
5↑
4↑
5 imm
4
6↑
5
7↑
6
7
11↑
8↑
10
9
11
John H. Dick

PLATE 72

PLATE 72

1. (1821) WHITECHEEKED TIT *Aegithalos leucogenys* R Sparrow-: 10 cm. Ilex and tamarisk scrub along rivers, etc. NWFP, Baluchistan. Seasonally between *c.* 450 and 3600 m.

2. (1822) WHITETHROATED TIT *Aegithalos niveogularis* R Sparrow-: 10 cm. Subalpine mixed deciduous and coniferous forest. W Himalayas, seasonally between *c.* 1800 and 3600 m.

3. (1823) RUFOUSFRONTED TIT *Aegithalos iouschistos* R Sparrow-; 10 cm. Shrubby undergrowth in conifer and mixed forest. E Himalayas, seasonlly between *c.* 2400 and 3600 m.

4. (1818, 1819) REDHEADED TIT *Aegithalos concinnus* R Sparrow-: 10 cm. Light open hill forest. From Hazara east along the Himalayas. Seasonally between *c.* 600 and 3600 m.

5. (1820) Ssp. *manipurensis* of 1818. Assam, Meghalaya, Nagaland, Manipur and Mizoram.

6. (1815) FIRECAPPED TIT *Cephalopyrus flammiceps* RM Sparrow-: 9 cm. Poplar, willow, oak and mixed forest; orchards. *Summer*: Himalayas, between *c.* 2000 and 3500 m. *Winter*: N and C India.

7. (1817) PENDULINE TIT *Remiz pendulinus* M Sparrow-: 10 cm. Tamarisk-Acacia scrub along rivers and canals. Pakistan and Punjab plains. Possibly breeds in Ladakh.

8. (EL) Blackheaded ssp. *nigricans* of 1817. Afghanistan, Iran.

9. (1808) BROWN CRESTED TIT *Parus dichrous* R Sparrow-: 12 cm. Mixed deciduous and conifer forest. Himalayas, from *c.* 2200 m to timber-line.

10. (1803) COAL TIT *Parus ater* R Sparrow-: 10 cm. Subalpine conifer forest. Himalayas from Garhwal east, seasonally *c.* 1800-3600 m.

11. (1804) BLACK TIT *Parus rufonuchalis* R Sparrow±: 13 cm. Mixed oak and conifer forest. N Baluchistan; W Himalayas. Seasonally from *c.* 500 m to timber-line.

12. (1802) CRESTED BLACK TIT *Parus melanolophus* R Sparrow-: 11 cm. Oak and conifer forest. W Himalayas, seasonally from *c.* 100 m to timber-line.

13. (1805) RUFOUSBELLIED CRESTED TIT *Parus rubidiventris* R Sparrow-: 10 cm. Subalpine conifer and rhododendron scrub. Himalayas from Tehri Garhwal east, seasonally between *c.* 2200 and 4200 m; NE hill states.

14. (1812) BLACKSPOTTED YELLOW TIT *Parus spilonotus* R Sparrow±: 14 cm. Light mixed forest. E Himalayas, seasonally between 1400 and 3700 m; NE hill states.

15. (1800) YELLOWBREASTED BLUE TIT, or AZURE TIT *Parus cyanus flavipectus* V Sparrow-: 13 cm. Willow, juniper and birch scrub in river beds, etc. Chitral.

16. (1800a) (TIEN SHAN) YELLOWBREASTED BLUE TIT *Parus cyanus tianshanicus* V Sparrow - : 13 cm. Bushes near water. Hunza.

17. (1799) GREENBACKED TIT *Parus monticolus* R Sparrow-: 13 cm. Hill forest. Himalayas, up to *c.* 3600 m; NE hill states.

18. (1794) GREY TIT *Parus major* R Sparrow-: 13 cm. Lightly wooded country. Subcontinent; Sri Lanka.

19. (1798) WHITEWINGED BLACK TIT *Parus nuchalis* R Sparrow-: 13 cm. Hummocky semi-desert country. Kutch, N Gujarat, Rajasthan. Disjunctly in E Ghats and Karnataka.

20. (1789) SULTAN TIT *Melanochlora sultanea* R Bulbul±: 20 cm. Foothills forest, up to *c.* 700 m. E Himalayas; NE hill states; Bangladesh.

21. (1814) YELLOWBROWED TIT *Sylviparus modestus* R Sparrow-: 9 cm. Mixed deciduous and conifer forest. Himalayas, NE hill states, seasonally between *c.* 1200 and 4200 m.

22. (1809) YELLOWCHEEKED TIT *Parus xanthogenys* R Sparrow±: 14 cm. Himalayan and peninsular forests. Subcontinent.

23. (EL) BLUE-GREY TIT *Parus bokarensis*. Russian Turkestan and Afghanistan.

PLATE 73

1. (1826) EUROPEAN NUTHATCH *Sitta europaea nagaensis* R Sparrow-: 12 cm. Mixed deciduous and evergreen, and coniferous forest. NE hill states, between 1400 and 2800 m.

2. (1834) WHITETAILED NUTHATCH *Sitta himalayensis* R sparrow-: 12 cm. Deciduous or evergreen broadleafed forest. Himalayas; NE hill states. Seasonally between *c.* 930 and 3400 m.

3. (1830) CHESTNUTBELLIED NUTHATCH *Sitta castanea* R Sparrow-: 12 cm. Deciduous forest and light woods. Lower Himalayas, up to *c.* 1800 m; practically all Subcontinent.

4. (1824) EUROPEAN NUTHATCH *Sitta europaea cashmirensis* R Sparrow-: 12 cm. Coniferous, deciduous or mixed forest. N Baluchistan; W Himalayas; up to 3300 m.

5. (1836) ROCK NUTHATCH *Sitta tephronota* R Sparrow±: 15 cm. Rocky valleys with broken cliffs, near streams. Baluchistan, seasonally between 300 and 2500 m.

6. (1832) WHITECHEEKED NUTHATCH *Sitta leucopsis* R Sparrow-: 12 cm. Pine, fir, deodar and mixed forest. Himalayas, seasonally from *c.* 1800 m to timber-line.

7. (1838) VELVETFRONTED NUTHATCH *Sitta frontalis* R Sparrow-: 10 cm. Moist-deciduous and evergreen hilly biotope. Lower Himalayas; practically all Subcontinent excepting NW; Sri Lanka.

8. (1837) BEAUTIFUL NUTHATCH *Sitta formosa* R Sparrow±: 15 cm. Deep, wet semi-evergreen and evergreen forest. E Himalayas; NE hill states. Seasonally between *c.* 330 and 2400 m.

9. (1839) WALL CREEPER *Tichodroma muraria* R Sparrow+: 17 cm. Gorges and vertical cliffs. High Himalayas, above 3300 m. *Winter*: Foothills; sometimes N Indian plains.

10. (1842) TREE CREEPER *Certhia familiaris* R Sparrow-: 12 cm. Pine, deodar, birch and rhododendron forest. Himalayas, seasonally from *c.* 1700 m up to timber-line.

11. (1847) HIMALAYAN TREE CREEPER *Certhia himalayana* R Sparrow-: 12 cm. Forest of pine, fir, deodar, spruce, etc. N Baluchistan; W Himalayas. Seasonally from the plains up to timber-line.

12. (1851) NEPAL TREE CREEPER *Certhia nipalensis* R Sparrow-: 12 cm. Oak and mixed deciduous and coniferous forest. E Himalayas seasonally between 1500 and 3500 m.

13. (1849) SIKKIM TREE CREEPER *Certhia discolor* R Sparrow-: 12 cm. Deciduous and oak and rhododendron forest. E Himalayas; NE hill states. Seasonally between *c.* 300 and 3600 m.

14. (1841) SPOTTED GREY CREEPER *Salpornis spilonotus* R Sparrow-: 13 cm. Open deciduous forest and groves. C and peninsular India; capricious.

15. (1235) BEARDED TIT-BABBLER, or REEDLING *Panurus biarmicus* V Sparrow±: 15 cm. Dense grass and tamarisk scrub. Pakistan (NWFP).

16. (1249) GREYHEADED PARROTBILL *Paradoxornis gularis* R Bulbul-: 16 cm. Bamboo, bushes and low trees. E Himalayas; NE hill states; up to *c.* 2400 m.

17. (1251) GOULD's or BLACKTHROATED PARROTBILL *Paradoxornis flavirostris* R Bulbul±: 19 cm. Elephant grass thickets, bamboo jungle. E Himalayas, NE hill states; Bangladesh; up to *c.* 2400 m.

18. (1252) WHITETHROATED PARROTBILL *Paradoxornis guttaticollis* R Bulbul±: 19 cm. Scrub and grass on abandoned cultivation, bushes and bamboo jungle. Patkai Range; Nagaland; Chittagong hill tracts (Bangladesh); between *c.* 900 and 2100 m.

19. (1247) GREATER REDHEADED PARROTBILL *Paradoxornis ruficeps* R Bulbul-: 18 cm. Wet forest, bamboo and scrub jungle, reed-beds, etc. E Himalayas; NE hill states; Bangladesh; up to *c.* 1800 m.

20. (1245) Ssp. *oatesi* of 1246. Sikkim; Darjeeling; *c.* 2100 m. Very rare.

21. (1246) LESSER REDHEADED SUTHORA or PARROTBILL *Paradoxornis atrosuperciliaris* R Sparrow±: 15 cm. Bamboo, high grass and scrub jungle. E Himalayas; NE hill states; Bangladesh. Seasonally from the foothills to 1500 m.

22. (1241) ORANGE SUTHORA *Paradoxornis nipalensis humii* R Sparrow-: 10 cm. Mixed deciduous and bamboo jungle. E Himalayas, between 1200 and 3300 m.

23. (1242) ORANGE SUTHORA *Paradoxornis nipalensis poliotis* R Sparrow-: 10 cm. Bamboo jungle and dense evergreen forest. E Himalayas; NE hill states: *c.* 600-2600 m.

24. (1238) FULVOUSFRONTED SUTHORA or PARROTBILL *Paradoxornis fulvifrons* R Sparrow-: 12 cm. Dense bamboo forest. E Himalayas, between *c.* 2700 and 3400 m.

PLATE 73
1
2
3 ♀
3 ♂
4 ♀
4 ♂
5
6
7
8
9 S
9 W
10
11
12
13
14
15 ♂
♀
16
17
18
19
20
21
22
23
24
John H. Dick

PLATE 74

PLATE 74

1. (1209) REDFRONTED BABBLER *Stachyris rufifrons* R Sparrow-: 12 cm. Dense undergrowth, thickly forested ravines, etc. E Himalayan foothills; NE hill states; Bangladesh; Orissa; E Ghats.

2. (1211) REDBILLED BABBLER *Stachyris pyrrhops* R Sparrow-: 10 cm. Light forest, bamboo and scrub jungle. W Himalayas, between *c.* 750 and 2400 m, occasionally 300 m.

3. (1212) GOLDHEADED BABBLER *Stachyris chrysaea* R Sparrow-: 10 cm. Evergreen biotope: dense undergrowth and secondary jungle. E Himalayas (800-2600 m); NE hill states (300-1800 m); Bangladesh (above *c.* 1200 m).

4. (1228) YELLOWBREASTED BABBLER *Macronous gularis* R Sparrow-: 11 cm. Evergreen and moist-deciduous biotope: foothills bamboo jungle, long grass and brushwood. E Himalayas; NE hill states; Bangladesh; eastern peninsular India.

5. (1210) REDHEADED BABBLER *Stachyris ruficeps* R Sparrow-: 12 cm. Dense evergreen undergrowth, bamboo and secondary jungle. E Himalayas; NE hill states; Bangladesh. Seasonally between *c.* 600 and 2700 m.

6. (1222) RUFOUSBELLIED BABBLER *Dumetia hyperythra* R Sparrow-: 13 cm. Deciduous or evergreen biotope; grassland and scrub. Base of Himalayas, Simla to Darjeeling duars, Bengal and Bangladesh, south to Krishna river, west to a line Hyderabad-Jalna-Mhow-Jhansi. Sri Lanka.

7. (1219) Ssp. *abuensis* of 1222. S Rajasthan, Kathiawar peninsula, E Gujarat, W Satpuras and W Ghats south to Pune, east to about Aurangabad; S India south to Krishna river and Mahabaleshwar; Sri Lanka.

8. (1218) AUSTEN'S SPOTTED BABBLER *Stachyris oglei* R Sparrow-: 13 cm. Dense evergreen scrub in hills and rocky ravines. Arunachal Pradesh (Lohit and Tirap districts), between 850 and 1800 m.

9. (1214) BLACKTHROATED BABBLER *Stachyris nigriceps* R Sparrow-: 12 cm. Evergreen secondary scrub and bamboo jungle. E Himalayas; NE hill states; Bangladesh; up to 1800 m.

10. (1224) BLACKHEADED BABBLER *Rhopocichla atriceps* R Sparrow-: 13 cm. Evergreen biotope: sholas, thickets, cane brakes in dank ravines, etc. Southern W Ghats complex.

11. (1226) Sri Lanka ssp. *siccatus* of 1224.

12. (1187) Ssp. *formosus* of 1186. Restricted to NE hill states, between *c.* 900 and 2400 m.

13. (1186) CORALBILLED SCIMITAR BABBLER *Pomatorhinus ferruginosus* R Myna-: 22 cm. Dense shrubbery and ringal bamboo jungle. E Himalayas; NE hill states; between *c.* 600 and 3800 m.

14. (1191) SLENDERBILLED SCIMITAR BABBLER *Xiphirhynchus superciliaris* R Bulbul±: 20 cm. Bushes on steep grassy hillsides, oak and rhododendron forest and bamboo thickets. E Himalayas; NE hill states; Chittagong hills. Seasonally from the foothills up to *c.* 3400 m.

15. (1185) LARGE SCIMITAR BABBLER *Pomatorhinus hypoleucos* R Myna±: 28 cm. Dense bamboo, cane and elephant grass jungle. E Himalayas; NE hill states; up to *c.* 1200 m. Bangladesh.

16. (1178) RUFOUSNECKED SCIMITAR BABBLER *Pomatorhinus ruficollis* R Bulbul±: 19 cm. Moist-deciduous forest and scrub jungle. Himalayas from Kumaon east; NE hill states; Bangladesh. Seasonally between *c.* 800 and 3300 m.

7. (1189) LONGBILLED SCIMITAR BABBLER *Pomatorhinus ochraceiceps* R Bulbul+: 23 cm. Dense evergreen and bamboo jungle. Mishmi Hills (Arunachal Pradesh); NE hill states; up to *c.* 2400 m.

18. (1169) SLATYHEADED SCIMITAR BABBLER *Pomatorhinus horsfieldi schisticeps* R Myna-: 22 cm. Ssp. of 1173. South India.

19. (1173) SLATYHEADED SCIMITAR BABBLER *Pomatorhinus horsfieldi* R Myna-: 22 cm. Evergreen and deciduous forest, thorn scrub and bamboo jungle. Himalayas; NE hill states up to *c.* 1500 m; practically entire Subcontinent (several other spp.). Sri Lanka.

20. (1181) RUSTYCHEEKED SCIMITAR BABBLER *Pomatorhinus erythrogenys* R Myna+: 25 cm. Scrubby forest edges, overgrown nullahs, bush-clad hillsides. Himalayas; NE hill states; Bangladesh; up to *c.* 2700 m.

21. (EL) SPOTBREASTED SCIMITAR BABBLER *Pomatorhinus erythrocnemis* Taiwan.

PLATE 75

1. (1770) WREN *Troglodytes troglodytes* R Sparrow-: 9 cm. Juniper scrub, rocks and boulders in birch, oak or rhododendron forest, etc. N Baluchistan; W Himalayas east to Garhwal. Seasonally between *c.* 1200 and 5000 m.

2. (1771) Ssp. *nipalensis* of 1770. From Tehri Garhwal east to Bhutan and Arunachal Pradesh.

3. (1199) BROWN, or LESSER SCALYBREASTED, WREN-BABBLER *Pnoepyga pusilla* R Sparrow-: 9 cm. Wet-evergreen forest with mossy boulders, etc. E Himalayas; NE hill states; Bangladesh (Chittagong region); up to 3000 m.

4. (1200) TAILED WREN-BABBLER *Spelaeornis caudatus* R Sparrow-: 9 cm. Damp undergrowth in thick evergreen forest. E Himalayas; NE hill states.

5. (1198) SCALYBREASTED WREN-BABBLER *Pnoepyga albiventer* R Sparrow-: 10 cm. Dense undergrowth in wet-evergreen forest. Himalayas from Dula Dhar east; NE hill states. Seasonally between *c.* 600 and 3900 m.

6. (1206) SPOTTED WREN-BABBLER *Spelaeornis formosus* R Sparrow-: 10 cm. Dank mossy rhododendron forest with thick fern ground-cover. E Himalayas; NE hill states; Bangladesh. Seasonally between *c.* 1200 and 2300 m.

7. (1203) STREAKED LONGTAILED WREN-BABBLER *Spelaeornis chocolatinus* R Sparrow-: 10 cm. Deep evergreen forest with undergrowth of bracken and ferns, etc. NE hill states, seasonally between *c.* 1200 and 2400 m.

8. (1201) MISHMI WREN *Spelaeornis badeigularis* R Sparrow-: 9 cm. Subtropical wet forest. Arunachal Pradesh (Lohit and Tirap districts).

9. (1193) LONGBILLED WREN-BABBLER *Rimator malacoptilus* R Sparrow-: 12 cm. Forest undergrowth and dense scrub in steep, broken country. E Himalayas; NE hill states; between 900 and 2700 m.

10. (1205) LONGTAILED SPOTTED WREN-BABBLER *Spelaeornis troglodytoides* R Sparrow-: 10 cm. Undergrowth and bamboo in wet temperate forest. E Bhutan.

11. (1195) SMALL WREN-BABBLER *Napothera epilepidota* R Sparrow-: 10 cm. Dense dark forest, E Himalayas; NE hill states; up to *c.* 1800 m.

12. (1194) STREAKED, or SHORT-TAILED, WREN-BABBLER *Napothera brevicaudata* R Sparrow-: 12 cm. Damp and shady hill forest with ravines and steep slopes. E Arunachal Pradesh; NE hill states; between *c.* 700 and 2100 m.

13. (1202) LONGTAILED WREN-BABBLER *Spelaeornis longicaudatus* R Sparrow-: 11 cm. Undergrowth in deep evergreen mossy forest. E Himalayas; NE hill states; between *c.* 1000 and 2000 m.

14. (1207) WEDGEBILLED WREN *Sphenocichla humei* R Bulbul-: 17-18 cm. Evergreen forest and dense jungle with large trees and bamboo. E Himalayas; NE hill states. Seasonally between *c.* 900 and 2300 m.

15. (1164) BROWN BABBLER *Pellorneum albiventre* R Sparrow±: 15 cm. Evergreen or moist-deciduous foothills scrub and bamboo jungle. E Himalayas; NE hill states; Bangladesh; up to *c.* 1500 m.

16. (1161) BROWNCAPPED BABBLER *Pellorneum fuscocapillum* R Bulbul-: 16 cm. Dense forest, scrub and undergrowth. Sri Lanka.

17. (1166) TICKELL'S BABBLER *Trichastoma tickelli* R Sparrow±: 15 cm. Deciduous or evergreen biotope; bamboo thickets and heavy scrub. Arunachal Pradesh; NE hill states; Bangladesh; up to *c.* 2100 m.

18. (1167) ABBOTT'S BABBLER *Trichastoma abbotti* R Bulbul-: 15 cm. Evergreen and deciduous biotope: tangled thickets in wet deep foothills jungle. E Himalayas; NE hill states; NW Bengal, Orissa, an isolated population in Andhra Pradesh; Bangladesh (Chittagong region); up to *c.* 600 m.

19. (1160) MARSH SPOTTED BABBLER *Pellorneum palustre* R Sparrow±: 15 cm. Marshy tree-jungle with bushes and high grass. Arunachal Pradesh; NE hill states; Bangladesh; up to *c.* 800 m.

20. (1154) SPOTTED BABBLER *Pellorneum ruficeps* R Bulbul-: 15 cm. Evergreen or deciduous foothills forest, scrub and secondary growth. Himalayas; NE hill states; most of Subcontinent.

1
3 ○
5 ○
4
2
3 ●
5 ●
8
6
7
9
10
12
11
13
14
15
16
17
19
18
20
John H. Dick

1
2
3
4
5
6
7
8
9
10
11
12
13
14
15
16
John H. Dick

PLATE 76

1. (1320) PLAINCOLOURED LAUGHING THRUSH *Garrulax subunicolor* R Myna±: 23 cm. Thickets of *Rubus*, dwarf rhododendron, bamboo, and undergrowth in mixed deciduous forest. E Himalayas, seasonally between *c.* 800 and 3900 m.

2. (1314) STREAKED LAUGHING THRUSH *Garrulax lineatus* R Bulbul±: 20 cm. Juniper scrub, bushes in open conifer forest and bush-covered slopes. N Baluchistan; Himalayas up to *c.* 3900 m.

3. (1326) Ssp. *nigrimentum* of 1324. E Himalayas, seasonally between *c.* 1200 and 3300.

4. (1324) REDHEADED LAUGHING THRUSH *Garrulax erythrocephalus* R Myna+: 28 cm. Damp forest with dense undergrowth, along nullahs, etc. Himalayas from Changla Gali and Murree east; NE hill states. Seasonally between *c.* 600 and 3300 m.

5. (1319) BLUEWINGED LAUGHING THRUSH *Garrulax squamatus* R Myna+: 25 cm. Humid dense hill bamboo thickets and rhododendron bushes along streams, etc. E Himalayas; NE hill states between *c.* 900 and 3400 m.

6. (1318) BROWNCAPPED LAUGHING THRUSH *Garrulax austeni* R Myna±: 22 cm. Oak and rhododendron forest, bushes in ravines and clearings, and bamboo thickets. NE hill states, between *c.* 1200 and 2700 m.

7. (1317) MANIPUR STREAKED LAUGHING THRUSH *Garrulax virgatus* R Myna+: 25 cm. Damp evergreen forest with heavy undergrowth. NE hill states, between *c.* 900 and 2400 m.

8. (1304) SPOTTEDBREASTED LAUGHING THRUSH *Garrulax merulinus* R Myna±: 22 cm. Deep damp forest with heavy undergrowth. NE hill states, between *c.* 900 and 2400 m.

9. (1294) RUFOUSCHINNED LAUGHING THRUSH *Garrulax rufogularis* R Myna±: 22 cm. Oak and rhododendron forest, scrub jungle and secondary growth in subtropical and wet-temperate biotope. Himalayan foothills from Murree east; NE hill states; Bangladesh; between *c.* 600 and 1900 m.

10. (1279) STRIATED LAUGHING THRUSH *Garrulux striatus* R Myna+: 28 cm. Dense forest with heavy undergrowth; also secondary and scrub jungle. Himalayas from Kulu east; NE hill states. Seasonally between *c.* 750 and 2850 m.

11. (1275) NECKLACED LAUGHING THRUSH *Garrulax moniligerus* R Myna+: 27 cm. Tropical evergreen and moist-deciduous forest and secondary jungle. E Himalayas; NE hill states; Bangladesh (Chittagong); up to *c.* 1000 m.

12. (1270) CHINESE BABAX *Babax lanceolatus* R Myna+: 28 cm. Thin scattered forest and open bracken-covered hillsides. Mizoram, above *c.* 1500 m.

13. (1277) BLACKGORGETED LAUGHING THRUSH *Garrulax pectoralis* R Myna+: 29 cm; Sal and pine forest, secondary growth and bamboo jungle. E Himalayas; NE hill states; Bangladesh (Chittagong region); up to *c.* 1700 m.

14. (1271) GIANT TIBETAN BABAX *Babax waddelli* R Pigeon±: 31 cm. Arid scrub. Tibetan facies of extreme NE Sikkim; Arunachal Pradesh(?); between *c.* 2800 and 4500 m.

15. (1297) GIANT LAUGHING THRUSH *Garrulax ocellatus maximus* R Pigeon+: 35 cm. Subalpine forest. SE Tibet, between 2200 and 2900 m.

16. (1299) WHITESPOTTED LAUGHING THRUSH *Garrulax ocellatus* R Pigeon±: 32 cm. Light forest with undergrowth, thick rhododendron scrub. Himalayas from Garhwal east, between *c.* 2100 and 4530 m.

PLATE 77

PLATE 77

1. (1331) CRIMSONWINGED LAUGHING THRUSH *Garrulax phoeniceus* R Myna±: 23 cm. Undergrowth in evergreen or shady deciduous forest, and dense secondary growth along streams. E Himalayas; NE hill states; Bangladesh; *c.* 900-2400 m.

2. (1307) NILGIRI LAUGHING THRUSH *Garrulax cachinnans* R Myna-: 20 cm. Dense undergrowth in sholas, hill station gardens, etc. Nilgiri Hills (S India), above *c.* 1200 m.

3. (1309) WHITEBREASTED LAUGHING THRUSH *Garrulax jerdoni* R Myna-: 20 cm. Evergreen biotope: thickets, scrub, secondary jungle, edges of sholas, hill station gardens, etc. Southern W Ghats complex in Kerala and Tamil Nadu, above *c.* 1200 m.

4. (1291) ASHY LAUGHING THRUSH *Garrulax cineraceus* R Myna± : 22 cm. Wet evergreen scrub and secondary growth in forest. Assam, Nagaland and Manipur; up to *c.* 2400 m.

5. (1286) YELLOWTHROATED LAUGHING THRUSH *Garrulax galbanus* R Myna±: 23 cm. Wet evergreen biotope: Open jungle and tall grass intermixed with trees and shrubs. S Assam; NE hill states; Bangladesh (Chittagong hill tracts); between *c.* 800 and 1800 m.

6. (1287) RUFOUSVENTED LAUGHING THRUSH *Garrulax delesserti* R Myna±: 23 cm. Dense evergreen undergrowth. Southern W Ghats complex up to *c.* 1500 m.

7. (1288) Ssp. *gularis* of 1287. Foothills of E Bhutan and Arunachal Pradesh; NE hill states; Bangladesh; up to *c.* 1800 m.

8. (1306) WHITEBROWED LAUGHING THRUSH *Garrulax sannio* R Myna±: 23 cm. Dense forest, scrub pastures, bamboo or secondary growth; also open hillsides under bracken and scrub. Assam, Nagaland and Manipur; between 1000 and 1800 m.

9. (1303) RUFOUSNECKED LAUGHING THRUSH *Garrulax ruficollis* R Myna±: 23 cm. Scrub and grass, bamboo jungle, secondary growth, etc. E Himalayas; NE hill states; Bangladesh (Chittagong hill tracts) up to *c.* 1500 m.

10. (1285) CHESTNUTBACKED LAUGHING THRUSH *Garrulax nuchalis* R Myna±: 23 cm. Dense wet evergreen scrub jungle on broken ground. Mishmi Hills; NE hill states; up to *c.* 900 m.

11. (1321) PRINCE HENRI'S LAUGHING THRUSH *Garrulax henrici* R Myna+: 25 cm. Scrub in the dry as well as wet zones. Adjacent SE Tibet, seasonally between *c.* 2000 and 4500 m.

12. (1322) BLACKFACED LAUGHING THRUSH *Garrulax affinis* R Myna+: 25 cm. Rhododendron bushes, scrub oak and bamboo in mixed oak and conifer forest and dwarf rhododendron. E Himalayas, seasonally between 1500 and 4500 m.

13. (1290) VARIEGATED LAUGHING THRUSH *Garrulax variegatus* R Myna±: 24 cm. Open forest of fir and birch with rhododendron and bamboo growth and miscellaneous dense jungle. W Himalayas, seasonally between *c.* 1200 and 4100 m.

14. (1300) GREYSIDED LAUGHING THRUSH *Garrulax caerulatus* R Myna+: 25 cm. Undergrowth in forest ringal bamboo and scrub-covered hillsides. E Himalayas; NE hill states. Seasonally between *c.* 600 and 2700 m.

15. (1274) WHITETHROATED LAUGHING THRUSH *Garrulax albogularis* R Myna+: 28 cm. Dense forest, light jungle and open scrubby hillsides. Himalayas (now absent in Pakistan); NE hill states. Seasonally between *c.* 450 and 3400 m.

16. (1283) WHITECRESTED LAUGHING THRUSH *Garrulax leucolophus* R Myna+: 28 cm. Wet evergreen forest with dense secondary scrub and bamboo jungle. Himalayas; NE hill states; Bangladesh (Chittagong region); up to *c.* 2100 m.

PLATE 78

1. (1347) WHITEHEADED SHRIKE-BABBLER *Gampsorhynchus rufulus* R Bulbul+: 23 cm. Evergreen biotope: secondary growth, bamboo, bush-and-grass jungle, and undergrowth in forest. E Himalayas; NE hill states; Bangladesh; up to *c.* 1200 m.

2. (1352) HOARY BARWING *Actinodura nipalensis* R Bulbul±: 20 cm. Mixed oak, conifer and rhododendron forest with undergrowth. E Himalayas; NE hill states; between 1800 and 3300 m.

3. (1388) RUFOUSHEADED TIT-BABBLER *Alcippe brunnea* R Sparrow-: 13 cm. Dense bushes, bracken, bamboo jungle and secondary growth. Arunachal Pradesh, Assam, and NE hill states; between *c.* 900 and 2400 m.

4. (1348) SPECTACLED BARWING *Actinodura egertoni* R Bulbul+: 23 cm. Dense secondary growth and mixed trees-and-scrub in evergreen forest. E Himayalas; NE hill states; up to *c.* 2400 m.

5. (1355) AUSTEN'S BARWING *Actinodura nipalensis waldeni* R Bulbul±: 20 cm. Mossy evergreen and mixed forest. Arunachal Pradesh, Assam and NE hill states. Seasonally between 1500 and 3300 m.

6. (1379) CHESTNUT-HEADED TIT-BABBLER *Alcippe castaneceps* R Sparrow-: 10 cm. Heavy evergreen undergrowth at the edge of forest and on abandoned clearings. E Himalayas; NE hill states; Bangladesh (Chittagong region); up to *c.* 3000 m.

7. (1378) DUSKY-GREEN or YELLOWTHROATED TIT-BABBLER *Alcippe cinerea* R Sparrow-: 10 cm. Glades, etc. in deep evergreen forest; also bamboo clumps and cut-over scrub. E Himalayas; NE hill states; Bangladesh (Chittagong hill tracts); between *c.* 1000 and 2100 m.

8. (1386) REDTHROATED TIT-BABBLER *Alcippe rufogularis* R Sparrow-: 12 cm. Evergreen biotope: bamboo jungle, secondary growth in forest. E Himalayan foothills and duars; NE hill states Bangladesh (Chittagong hill tracts); up to *c.* 900 m.

9. (1392) NEPAL BABBLER *Alcippe nipalensis* R Sparrow-: 12 cm. Tropical wet evergreen biotope: forest, cut-over scrub, bamboo clumps, etc. E Himalayas; NE hill states; Bangladesh (Chittagong hill tracts); up to *c.* 1500 m.

10. (1385) GREYHEADED TIT-BABBLER *Alcippe cinereiceps* R Sparrow-: 11 cm. Secondary scrub jungle, rhododendron and dense bamboo forest. Assam, Nagaland and Manipur to Noa Dihing drainage (in Arunachal Pradesh). Seasonally between *c.* 1500 and 2800 m.

11. (1390) QUAKER BABBLER *Alcippe poioicephala* R Sparrow±: 15 cm. Evergreen and moist-deciduous biotope: forest, bamboo jungle, open scrub, canebrakes, etc. Peninsular India, hills and plains; Assam; NE hill states; Bangladesh.

12. (1375) WHITEBELLIED YUHINA *Yuhina zantholeuca* R Sparrow-: 11 cm. Evergreen or moist-deciduous biotope: glades or along streams, etc. Himalayas; NE hill states; Bangladesh (Chittagong hill tracts); up to *c.* 2600 m.

13. (1365) WHITEBROWED YUHINA *Yuhina castaniceps* R Sparrow-: 13 cm. Scrub and undergrowth in evergreen or light deciduous forest. E Himalayas; NE hill states; Bangladesh (Chittagong); up to *c.* 1500 m.

14. (1381) WHITEBROWED TIT-BABBLER *Alcippe vinipectus* R Sparrow-: 11 cm. Low scrub, light conifer forest, and forest edges with ample undergrowth and ringal bamboo. Himalayas; NE hill states. Seasonally between *c.* 1500 and 4200 m.

15. (1368) YELLOWNAPED YUHINA *Yuhina flavicollis* R Sparrow-: 13 cm. Oak and deciduous forest and open secondary jungle. Himalayas; NE hill states; Bangladesh (Chittagong hill tracts); up to *c.* 3000 m.

16. (1374) BLACKCHINNED YUHINA *Yuhina nigrimenta* R Sparrow-: 11 cm. Evergreen forest and secondary jungle. Himalayan foothills from Garhwal east; NE hill states; Bangladesh (Chittagong region) up to *c.* 1800 m.

17. (1373) RUFOUSVENTED YUHINA *Yuhina occipitalis* R Sparrow-: 13 cm. Evergreen forest, especially rhododendron and oak. E Himalayas, seasonally between 1500 and 3900 m.

18. (1372) STRIPETHROATED YUHINA *Yuhina gularis* R Sparrow±; 14 cm. Forest of oak, rhododendron or mixed with conifers. Himalayas from Garhwal east; NE hill states. Seasonally between *c.* 1800 and 3600 m.

19. (1366) WHITENAPED YUHINA *Yuhina bakeri* R Sparrow-: 13 cm. Secondary jungle and evergreen forest. E Himalayas; NE hill states; Bangladesh; up to *c.* 3000 m.

PLATE 78

1
2
3
4
5
6
7
8
9
10
11
12
13
14
15
John H. Dick

PLATE 79

1. (1229) REDCAPPED BABBLER *Timalia pileata* R Sparrow+: 17 cm. Wet-evergreen biotope: flood plains with tall grass, reed-beds, brushwood and scrub jungle. E Himalayan terai, duars, etc.; NE hill states; Bangladesh.

2. (1231) YELLOWEYED BABBLER *Chrysomma sinense* R Bulbul-: 18 cm. Scrub, thorn jungle, reed-beds, bamboo, etc. in dry-deciduous facies. Practically throughout the Subcontinent; Sri Lanka.

3. (1233) JERDON'S BABBLER *Chrysomma altirostre* R Bulbul-: 17 cm. Reed and elephant-grass jungle along rivers. Flood plains of the Indus in Pakistan; Bengal and Bhutan duars; plains of the Brahmaputra system in NE India; Bangladesh.

4. (1254) COMMON BABBLER *Turdoides caudatus* R Bulbul+: 23 cm. Thorn scrub jungle, rambling compounds, etc. in semi-desert and dry plains. Most of the Subcontinent; Lakshadweep. Absent in Sri Lanka.

5. (1256) STRIATED BABBLER *Turdoides earlei* R Bulbul±: 20 cm. Tall grass, reeds and bushes. Flood plains of the Indus, Ganges, Brahmaputra and other N Indian river systems.

6. (1269) SPINY BABBLER *Turdoides nipalensis* R Bulbul+: 25 cm. Dense secondary scrub on hillsides. W Nepal east to east-central Nepal; seasonally 900 to 2000 m.

7. (1265) JUNGLE BABBLER *Turdoides striatus* R Myna+: 25 cm. Deciduous forest, bamboo, cultivation and scrub jungle, etc. Subcontinent. Absent in Sri Lanka and Andaman and Nicobar Is.

8. (1258) LARGE GREY BABBLER *Turdoides malcolmi* R Myna+: 28 cm. Dry, open, sparsely scrubbed country, thorn jungle, and around cultivation in such. Continental and peninsular India.

9. (1267) WHITEHEADED BABBLER *Turdoides affinis* R Bulbul+: 23 cm. Secondary jungle, wooded compounds and around cultivation and villages. Southern peninsular India; Sri Lanka.

10. (1257) SLENDERBILLED BABBLER *Turdoides longirostris* R Bulbul+: 23 cm. Long grass, usually near water: locally in terai, duars and lowlands. NE India; Bangladesh.

11. (1266) CEYLON RUFOUS BABBLER *Turdoides rufescens* R (endemic) Myna+: 25 cm. Deep primeval forest, thickets and bamboo scrub in well-wooded areas. Sri Lanka.

12. (1259) RUFOUS BABBLER *Turdoides subrufus* R Myna+: 25 cm. Wet-evergreen and moist-deciduous biotope: dense scrub intermixed with tall grass, and bamboo brakes, etc. W Ghats complex from Mahabaleshwar south to Kerala and adjacent Tamil Nadu. Sri Lanka (absent).

13. (1237) BROWN SUTHORA or PARROTBILL *Paradoxornis unicolor* R Bulbul±: 21 cm. Exclusively in *maling* bamboo and dwarf rhododendron growth. E Himalayas, seasonally between *c.* 2000 and 3400 m.

14. (1272) ASHYHEADED LAUGHING THRUSH *Garrulax cinereifrons* R (endemic) Myna±: 23 cm. Dense, humid forest. Sri Lanka.

15. (1236) GREAT PARROTBILL *Canostoma aemodium* R Myna+: 28 cm. Ringal bamboo and rhododendron bushes. E Himalayas, between *c.* 2000 and 3600 m.

PLATE 80

1. (1333) SILVEREARED MESIA *Leiothrix argentauris* R Sparrow±: 15 cm. Evergreen biotope; open forest, scrub jungle, secondary growth. Himalayas from Garhwal east; NE hill states; Bangladesh (Chittagong region) up to *c.* 2100 m.

2. (1336) REDBILLED LEIOTHRIX *Leiothrix lutea* R Sparrow-: 13 cm. Scrub and secondary growth. Himalayas; NE hill states. Seasonally between *c.* 600 and 2700 m.

3. (1339) NEPAL CUTIA *Cutia nipalensis* R Bulbul±: 20 cm. Heavy oak and mossy evergreen forest. Himalayas from Kumaon east; NE hill states; between *c.* 900 and 2500 m.

4. (1340) RUFOUSBELLIED SHRIKE-BABBLER *Pteruthius rufiventer* R Bulbul-: 17 cm. Dense, moss-covered oak and evergreen forest. E Himalayas; NE hill states; between *c.* 1500 and 2700 m.

5. (1343) GREEN SHRIKE-BABBLER *Pteruthius xanthochlorus* R Sparrow-: 13 cm. Oak, spruce, hemlock and deodar forest and undergrowth. Himalayas; NE hill states. Seasonally between *c.* 1200 and 3600 m.

6. (1346) CHESTNUTFRONTED SHRIKE-BABBLER *Pteruthius aenobarbus* R Sparrow-: 11 cm. Open forest and edges of evergreen. Known only from the Garo Hills, Meghalaya (NE hill states).

7. (1345) CHESTNUT-THROATED SHRIKE-BABBLER *Pteruthius melanotis* R Sparrow-: 11 cm. Humid but cool, deep evergreen forest. E Himalayas; NE hill states; Bangladesh (Chittagong hill tracts). Seasonally between *c.* 700 and 2700 m.

8. (1341) REDWINGED SHRIKE-BABBLER *Pteruthius flaviscapis* R Myna-: 16 cm. Heavy forest of oak, rhododendron, etc. Himalayas; NE hill states. Seasonally between *c.* 1200 and 2700 m.

9. (1359) BARTHROATED SIVA *Minla strigula* R Sparrow±: 14 cm. Open mixed forest of birch, willow, oak, rhododendron scrub, bamboo. Himalayas; NE hill states. Seasonally between *c.* 800 and 3700 m.

10. (1362) BLUEWINGED SIVA *Minla cyanouroptera* R Sparrow±: 15 cm. Mixed deciduous and evergreen forest and secondary growth. Himalayas from Naini Tal east; NE hill states; Bangladesh (Chittagong region). Seasonally between *c.* 600 and 2500 m.

11. (1376) GOLDENBREASTED TIT-BABBLER *Alcippe chrysotis* R Sparrow-: 11 cm. Dense bamboo jungle, etc. on steep hillsides. E. Himalayas; NE hill states. Seasonally between 1800 and 3000 m.

12. (1357) REDTAILED MINLA *Minla ignotincta* R Sparrow±: 14 cm. Humid, dense deciduous or evergreen forest. E Himalayas; NE hill states; Bangladesh (Chittagong region); up to *c.* 3400 m.

13. (1396) BLACKCAPPED SIBIA *Heterophasia capistrata* R Bulbul+: 21 cm. Evergreen or moist-deciduous forest of oak, chestnut, fir, etc., Himalayas, up to *c.* 3500 m.

14. (1395) CHESTNUTBACKED SIBIA *Heterophasia annectens* R Bulbul-: 18 cm. Dense, humid evergreen forest. E Himalayas; NE hill states; up to *c.* 2300 m.

15. (1400) BEAUTIFUL SIBIA *Heterophasia pulchella* R Bulbul+: 22 cm. Subtropical wet forest. Arunachal Pradesh; NE hill states.

16. (1401) LONGTAILED SIBIA *Heterophasia picaoides* R Bulbul±, with a long tail: 30 cm. Open scrub with large trees or clearings in evergreen forest. E Himalayas; NE hill states; up to *c.* 900 m.

17. (1399) GREY SIBIA *Heterophasia gracilis* R Bulbul+: 21 cm. Deciduous pine or evergreen primeval forest. Meghalaya, the hills of Assam, the Patkai range, Nagaland and Manipur.

1 ♂
2 ♂
3 ♂
♀
4 ♂
♀
5
6 ♂
♀
7 ♂
♀
8 ♂
♀
9
10
11
12
13
14
15
16
17
John H. Dick

PLATE 81
1
♀
2 ♂
♀
3 ♂
4
♀
5 ♂
7 ♀
8 ♀
7 ♂
6
8 ♂
♀
9 ♂
♀
10 ♂
♀
11 ♂
♀
12 ♂
♀
13 ♂
♀
14 ♂
♀
15 ♂
♀
16 ♂
♀
17 ♂
John H. Dick

PLATE 81

1. (1635) GOULD'S SHORTWING *Brachypteryx stellata* R Sparrow-: 13 cm. Dense dwarf rhododendron and ringal bamboo on hillsides, and bare rocks and boulders in the alpine zone. Himalayas from Kumaon east; seasonally between *c.* 540 and 4200 m.

2. (1636) RUSTYBELLIED SHORTWING *Brachypteryx hyperythra* R Sparrow-: 13 cm. Undergrowth of shrubs, reeds and ringal bamboo. E Himalayas; NE hill states; up to *c.* 2900 m.

3. (1639) LESSER SHORTWING *Brachypteryx leucophrys* R Sparrow-: 13 cm. Dense undergrowth in humid forest and secondary growth near streams. Himalayas from Garhwal east; NE hill states; *c.* 900-3900 m.

4. (1637) RUFOUSBELLIED SHORTWING *Brachypteryx major* R Sparrow±: 15 cm. Well-wooded sholas, evergreen forest. Southern W Ghats complex of hills.

5. (1640) WHITEBROWED SHORTWING *Brachypteryx montana* R Sparrow-: 13 cm. Damp, shady oak and rhododendron forest, and dense brushwood in ravines or near streams. Himalayas from Garhwal east; NE hill states. Seasonally between *c.* 300 and 3600 m.

6. (1642) NIGHTINGALE *Erithacus megarhynchos* M Bulbul-: 18 cm. Bushes and gardens. Baluchistan; has strayed to Oudh terai.

7. (1643) RUBYTHROAT *Erithacus calliope* M Sparrow±: 15 cm. Dense scrub near water, long grass, sugarcane; occasionally tea gardens. NE and E India including Bangladesh, south to the Godavari delta (winter). Isolated examples from the Satpuras, Delhi and Bharatpur.

8. (1647) HIMALAYAN RUBYTHROAT *Erithacus pectoralis* RM Sparrow±: 15 cm. Dwarf rhododendron, juniper and other scrub, grassy hillsides, marshes and cultivation. *Summer*: Himalayas. *Winter*: Duars and lowland: N and NE India; Bangladesh. One record each from Uttar Pradesh and Karnataka. Seasonally from the foothills up to *c.* 4500 m.

9. (1644) BLUETHROAT *Erithacus svecicus* RM Sparrow±: 15 cm. Sugarcane, cotton fields, grass jungle, reeds, bushes near water, gardens and fallow fields. *Summer*: Gilgit, Baltistan Ladakh, N Kashmir and Spiti. *Winter*: Subcontinent, Andamans; Sri Lanka.

10. (1652) FIRETHROAT *Erithacus pectardens* V Sparrow±: 15 cm. Single record, Garo Hills, Meghalaya.

11. (1650) BLUE CHAT *Erithacus brunneus* RM Sparrow±: 15 cm. Evergreen biotope: dense undergrowth in open oak, conifer or shola forest. *Summer*: Himalayas; NE hill states; *c.* 1600-3300 m. *Winter*: E Himalayan foothills; SW India (W Ghats complex). E Ghats, Karnataka and Tamil Nadu (passage). Sri Lanka.

12. (1653) SIBERIAN BLUE CHAT *Erithacus cyane* V Sparrow±: 15 cm. Isolated records from Bengal duars, Manipur and S Andamans.

13. (1659) WHITEBROWED BUSH ROBIN *Erithacus indicus* R Sparrow±: 15 cm. Mixed subalpine forest of birch, fir, rhododendron and bamboo. Himalayas from Garhwal east; NE hill states. Seasonally from the foothills up to *c.* 4200 m.

14. (1656) ORANGEFLANKED BUSH ROBIN *Erithacus cyanurus* RM Sparrow±: 15 cm. Himalayas; Assam, Meghalaya and Manipur (winter). Seasonally from the foothills up to *c.* 4400 m.

15. (1658) GOLDEN BUSH ROBIN *Erithacus chrysaeus* R Sparrow±: 15 cm. Evergreen biotope: dense undergrowth in rhododendron, juniper and birch forest. Himalayas; NE hill states. Seasonally from the foothills up to *c.* 4600 m.

16. (EL) WHITETHROATED ROBIN *Irania gutturalis*. Iran, Afghanistan, etc.

17. (1660) RUFOUSBELLIED BUSH ROBIN *Erithacus hyperythrus* R? Sparrow-: 12 cm. Dwarf rhododendron, and bushes in open forest near streams. E Himalayas; NE hill states (winter only?). Seasonally from the foothills up to *c.* 3800 m.

PLATE 82

1. (1641) RUFOUS CHAT *Erythropygia galactotes* RM Bulbul-: 18 cm. Dry scrub jungle, tamarisks and stony, broken country. *Breeding*: Baluchistan in extreme border regions. *Winter* (or on passage): Pakistan and NW India. Sri Lanka (once).

2. (1669) EVERSMANN'S REDSTART *Phoenicurus erythronotus* M Sparrow±: 15 cm. Arid country — wasteland, scrub jungle, dry river beds, etc. W Himalayas east to Nepal; base to 2800 m.

3. (1672) E ssp.
4. (1671) W ssp.] BLACK REDSTART *Phoenicurus ochruros* RM Sparrow±: 15 cm. Stony, sparsely scrubbed broken country and near cultivation and villages. *Breeding* Himalayas, *c.* 2400-5200 m. *Winter*: Subcontinent. Absent in Sri Lanka and Andaman and Nicobar Is.

5. 1670) BLUEHEADED REDSTART *Phoenicurus caeruleocephalus* R Sparrow±: 15 cm. Rocky ground and steep rocky hillsides, juniper and open pine forest. Himalayas, seasonally from the foothills up to *c.* 3900 m.

6. (1673) REDSTART *Phoenicurus phoenicurus* M Sparrow±: 15 cm. Gardens, well-wooded areas. Baluchistan and Chitral.

7. (1674) HODGSON'S REDSTART *Phoenicurus hodgsoni* M Sparrow±: 15 cm. Dry or partly dry river-beds in forest, cultivation, scrub, or grassland. *Winter* Lower Himalayas from Naini Tal (U.P.) east; Assam; Manipur; Nagaland.

8. (1675) BLUEFRONTED REDSTART *Phoenicurus frontalis* R Sparrow±: 15 cm. Dwarf rhododendron; juniper and other scrub, and boulder-strewn slopes in the alpine zone. Himalayas. *Winter*: NE hill states; Bangladesh (Chittagong region). Seasonally between *c.* 1000 and 5300 m.

9. (1676) WHITETHROATED REDSTART *Phoenicurus schisticeps* R Sparrow±: 15 cm. Open, park-like forest, scrub in semi-dry areas. E Himalayas, 1400-4500 m.

10. (1678) GULDENSTADT'S REDSTART *Phoenicurus erythrogaster* R Sparrow+: 16 cm. Boulder-strewn meadows and slopes in dry, barren alpine country. Himalayas, 1500-5200 m.

11. (1677) DAURIAN REDSTART *Phoenicurus auroreus* M Sparrow±: 15 cm. Open forest, valley floors, cultivation and trees around upland villages. *Summer*: N Arunachal Pradesh 2800-3700 m. *Winter*: E Himalayan foothills: NE hill states; Bangladesh.

12. (1679) PLUMBEOUS REDSTART *Rhyacornis fuliginosus* R Sparrow-: 12 cm. Rushing torrents and streams. Himalayas; NE hill states. Foothills to *c.* 4400 m.

13. (1680) HODGSON'S SHORTWING *Hodgsonius phoenicuroides* R Bulbul±: 19 cm. Dense thickets of birch and juniper, bush jungle, undergrowth and forest edges. Himalayas, foothills to *c.* 4400 m.

14. (1681) WHITETAILED BLUE ROBIN *Cinclidium leucurum* R Bulbul -: 17 cm. Undergrowth in shady broadleafed forest, usually near streams. E Himalayas; NE hill states. Foothills to *c.* 2700 m.

15. (1682) BLUEFRONTED ROBIN *Cinclidium frontale* R Bulbul±: 19 cm. Heavy subtropical wet forest. E Himalayas; (E Nepal, Sikkim and Darjeeling). Foothills to at least 2250 m.

16. (1720) INDIAN ROBIN *Saxicoloides fulicata* R Sparrow±: 16 cm. Arid and stony country, semi-desert with scattered bushes, cultivations, and around habitations. Subcontinent; Sri Lanka.

17 (1717) Brown backed ssp. *cambaiensis* of 1720. Himalayan foothills and plains of Subcontinent south to Tapi river, east to Bengal.

18. (1692) BROWN ROCK CHAT *Cercomela fusca* R Sparrow+: 17 cm. Rocky hills, quarries, ruins, bungalows within town limits and suburbs. NW, N and C India; locally up to *c.* 1300 m.

19. (1716) WHITECAPPED REDSTART or RIVER CHAT *Chaimarrornis leucocephalus* R Bulbul±: 19 cm. Rocky swift-flowing mountain streams. Himalayas; NE hill states. Foothills to *c.* 5100 m.

PLATE 82
1
1
♀
2 ♂
3 ♂
3 ♂ W
3 ♀
♀
4 ♂
♀
5 ♂
♀
6 ♂
♀
7 ♂
♀
8 ♂
♀
9 ♂
♀
11 ♂
♀
10 ♂
♀
12 ♂
♀
13 ♂
♀
14 ♂
♀
♀
15 ♂
17 ♂
16 ♂
19
18
John H. Dick

PLATE 83
4 ♂
♀
3
1 ♂
♀
3 ♂
♀
2
5
7
6
9
8
11 ♀
11 ♂
10 ♂
12 ♀
12 ♂
14
13
John H. Dick

PLATE 83

1. (1665) SHAMA *Copsychus malabaricus* R Bulbul±, with long tail: overall 25 cm. Dry- and moist-deciduous secondary forest, hilly bamboo jungle, etc. Himalayas from Kumaon east; duars to *c.* 500 m. Subcontinent; Sri Lanka.

2. (1668) Ssp. *albiventris* of 1665. Andaman Is.

3. (1683) HODGSON'S GRANDALA *Grandala coelicolor* R Bulbul+: 23 cm. Rocky slopes, boulder-strewn alpine meadows, screes and cliffs. High Himalayas from Kashmir east; *c.* 2200-5400 m.

4. (1661) MAGPIE-ROBIN *Copsychus saularis* R Bulbul±: 20 cm. Dry- and moist-deciduous forest, secondary jungle, and near human habitations. Subcontinent, Andamans; Sri Lanka.

5. (1684) LITTLE FORKTAIL *Enicurus scouleri* R Sparrow-: 12 cm. Rocky streams and waterfalls. Himalayas; NE hill states; Bangladesh (Chittagong region); *c.* 300-3700 m.

6. (1686) SLATYBACKED FORKTAIL *Enicurus schistaceus* R Bulbul±, with long tail: overall 25 cm. Rocky hill torrents and forest streams. Himalayas from Kumaon east; NE hill states; Bangladesh. Base of hills to *c.* 1600 m.

7. (1685) BLACKBACKED FORKTAIL *Enicurus immaculatus* R Bulbul±, with long tail: overall 25 cm. Rocky mountain streams and banks of fast-flowing rivulets in dense forest. Himalayas from Garhwal east; NE hill states; Bangladesh. Base of hills to *c.* 1450 m.

8. (1688) SPOTTED FORKTAIL *Enicurus maculatus* R Bulbul±, with long tail: overall 25 cm. Boulder-strewn shady nullahs and streams in forest, and narrow gorges. Himalayas;; NE hill states; Bangladesh (Chittagong region); *c.* 460-3600 m.

9. (1687) LESCHENAULT'S FORKTAIL *Enicurus leschenaulti* R Myna±, with long tail: overall 28 cm. Cascading torrents and rocky rivulets in dense evergreen forest. E Himalayas; NE hill states; Bangladesh. Base of hills to *c.* 600 m.

10. (1691) GREEN COCHOA *Cochoa viridis* R Myna+: 28 cm. Dense evergreen forest watered by small streams, usually on precipitous grounds. Himalayas from Kumaon east; NE hill states; *c.* 700-1500 m.

11. (1690) PURPLE COCHOA *Cochoa purpurea* R Myna+: 28 cm. Dense humid evergreen forest, and undergrowth in ravines in pine forest. E Himalayas; NE hill states; *c.* 1000-3000 m.

12. (1727) CEYLON WHISTLING THRUSH *Myiophonus blighi* R (endemic) Myna-: 20 cm. Damp, heavy forest, fern-clad ravines. Sri Lanka.

13. (1728) MALABAR WHISTLING THRUSH *Myiophonus horsfieldii* R Myna+: 25 cm. Waterfalls and swift rocky hill streams in shady evergreen jungle. The hills of W and peninsular India including the E & W Ghats complex. Foothills to *c.* 2200 m.

14. (1729) BLUE WHISTLING THRUSH *Myiophonus caeruleus* R Pigeon±: 33 cm. Rivers and torrents in heavy forest, wooded ravines and gorges. N Baluchistan; Himalayas; NE hill states. Foothills to *c.* 4200 m.

PLATE 84

1. (1693) STOLICZKA'S BUSH CHAT *Saxicola macrorhyncha* R Sparrow±: 15 cm. Desert with scattered shrubs. Formerly Pakistan east of the Indus, Haryana and Uttar Pradesh; now believed to be confined to E Rajasthan and Gujarat.

2. (1697) STONE CHAT or COLLARED BUSH CHAT *Saxicola torquata* RM Sparrow-: 13 cm. Dry, scrub-covered hillsides, wasteland, fallow fields, tamarisk jungle, etc. Subcontinent excepting extreme southern peninsula and Sri Lanka; Andamans.

3. (1696) Mostly migrant ssp. *przewalskii* of 1697. Himalayan foothills from Kangra east to Arunachal Pradesh, and south through Assam, Meghalaya and Nagaland (winter). *Breeding*: Nepal in the Dolpo region between 2745 and 4515 m.

4. (1694) HODGSON'S BUSH CHAT *Saxicola insignis* M Sparrow+: 17 cm. Tall grassland, cane fields, reeds and tamarisks along dry river-beds. Gangetic plain; Nepal terai; Sikkim; Jalpaiguri duars.

5. (1700) PIED BUSH CHAT *Saxicola caprata* R Sparrow-: 13 cm. Cultivated fields, sparsely scrubbed hillsides, tamarisk, reeds, and coarse grass near water and cultivation. Subcontinent. Absent in Andaman and Nicobar Is.

6. (1704) JERDON'S BUSH CHAT *Saxicola jerdoni* R Sparrow±: 15 cm. Vast expanses of elephant grass. NE India; Bangladesh. Foothills to *c.* 700 m.

7. (1699) WHITETAILED STONE CHAT *Saxicola leucura* R Sparrow-: 13 cm. High grass, reeds and tamarisks, especially near large rivers. Sub-Himalayan terai and flood plains of the Indus, Ganges and Brahmaputra river systems; Bangladesh.

8. (1705) DARK-GREY BUSH CHAT *Saxicola ferrea* R Sparrow±: 15 cm. Open scrub-covered hillsides, forest edges, grasslands and cultivated country. *Breeding*: Himalayas; NE hill states; *c.* 1500-3300 m. *Winter*: Foothills, Gangetic plain, Madhya Pradesh, Bangladesh and NE India.

9. (1708) WHEATEAR *Oenanthe oenanthe* MV Sparrow±: 15 cm. Stony country and cultivation. On autumn passage through Pakistan and Kashmir. Madhya Pradesh (once).

10. (1706) ISABELLINE CHAT *Oenanthe isabellina* RM Sparrow+: 16 cm. Sandy semi-desert and wasteland with sparse bushes. *Breeding*: Baluchistan and NWFP. *Winter*: chiefly Pakistan and NW India (Rajasthan, Kutch, etc.). Sporadically elsewhere.

11. (1707) REDTAILED CHAT *Oenanthe xanthoprymna* R Sparrow+: 16 cm. Flat, stony and sandy semi-desert with sparse bushes; arid rocky slopes near small streams. *Breeding*: N Baluchistan. *Winter*: Pakistan; Rajasthan, Gujarat, etc.

12. (1710) DESERT WHEATEAR *Oenanthe deserti* RM Sparrow±: 15 cm. Arid, semi-desert broken tracts, and canal-irrigated desert cultivation. *Breeding*: N Baluchistan, Kashmir east to NW Nepal. *Winter*: Pakistan, N and C and peninsular India; Sri Lanka.

13. (1715) PLESCHANKA'S PIED CHAT or WHEATEAR *Oenanthe pleschanka* M Sparrow±: 15 cm. Stony wasteland. NWFP and Gilgit east through Ladakh to Lahul.

14. (1714) HUME'S CHAT *Oenanthe alboniger* R Sparrow+: 17 cm. Steep-sided boulder-strewn nullahs. Baluchistan from Makran coast, Sind, Baltistan, Astor to Gilgit.

15. (1711) BARNES'S CHAT *Oenanthe finschi* R Sparrow+: 17 cm. Dry stony foothills, and arid semi-desert plains. N Baluchistan.

16. (1713) HOODED CHAT *Oenanthe monacha* R? Sparrow+: 17 cm. Desolate ravines, barren stony slopes bordering streams. *Winter*: Makran coast, Pakistan. *Breeding*: probably in the Soorjana hills, Las Bela.

17. (1712) PIED CHAT *Oenanthe picata* R Sparrow+: 17 cm. Boulder-strewn barren country, sparsely scrubbed sand-dunes, ravines, cultivation, outskirts of desert villages, etc. *Breeding*: From Baluchistan, Chitral, Baltistan to Gilgit. *Winter*: Pakistan; NW and C India. 17a. *picata* morph; 17b. *capistrata* morph; 17c. *opistholeuca* morph.

1♂ S
♀
2♂S
4♂ S
3♂W
♀
3♂ S
♀
5♂
♀
7♂
♀
6♂
♀
8♂
♀
9♂S
10↓
10
11
♀
12♂S
12♂W
♀
15♂ W
15♂ S
♀
13♂
14♂
17 a♂
a♀
17b♂
17 c♂
b♀
c♀
♀
16♂
♀
17♂
John H. Dick

PLATE 85
1 ♂
♀
3
4 ♂
♀
2
5 ♂
♀
7 ♂
♀
6 ♂
♀
8 ♂
♀
9 ♂
♀
10 ♂
11 ♂
12 ♀
14
13 ♂
♀
17 ♂
15 ♂
♀
16 ♂
♀
John H. Dick

PLATE 85

1. (1724) CHESTNUTBELLIED ROCK THRUSH *Monticola rufiventris* R Myna±: 24 cm. Open forest of pine, oak, fir and deodar on steep hillsides. Himalayas; NE hill states. *Breeding*: *c.* 1200-3300 m. *Winter*: Foothills, occasionally plains, Bangladesh.

2. (1733) ORANGEHEADED GROUND THRUSH *Zoothera citrina citrina*]
3. (1734) Whitethroated spp. of 1733. *Z. c. cyanotus*] M Myna-: 21 cm. Damp forest with plentiful undergrowth, mixed secondary and bamboo jungle, and groves near habitations. *Summer* (breeding): Himalayas; NE India including Bangladesh; foothills to *c.* 2300 m (2). Peninsular Indian hills (3). *Winter*: the Subcontinent locally; Sri Lanka; Andaman and Nicobar Is.

4. (1726) BLUE ROCK THRUSH *Monticola solitarius* RM Bulbul+: 23 cm. Barren rocky country, cliffs and rocks along seashore, quarries, old forts and buildings. *Summer* (breeding): Baluchistan; W Himalayas, 1500-3000 m. *Winter*: From Himalayan foothills south throughout Subcontinent including Sri Lanka, Andamans and Nicobars.

5. (1723) BLUEHEADED ROCK THRUSH *Monticola cinclorhynchus* RM Bulbul-: 17 cm. *Summer* (breeding): Open forest and rocky grass-covered slopes. Himalayas; NE hill states. *Winter*: Evergreen and moist-deciduous secondary jungle. W Ghats complex from the Tapti river south; E Ghats and Assam hills.

6. (1748) TICKELL'S THRUSH *Turdus unicolor* R Myna-: 21 cm. Open deciduous forest, groves, orchards, gardens. *Summer* (breeding) Himalayas, 1200-270 m. *Winter*: From Himachal Pradesh east along the foothills; locally elsewhere in the peninsula. Bangladesh.

7. (1722) ROCK THRUSH *Monticola saxatilis* RM Bulbul±: 19 cm. Rocky hillsides. *Breeding*: N Baluchistan *c.* 2000-3000 m. In passage through NWFP. Ladakh, Kashmir and the hills of Sind.

8. (1750) GREYWINGED BLACKBIRD *Turdus boulboul* R Myna+: 28 cm. Humid broadleafed forest of oak, rhododendron, etc.; woods, bush jungle and village precincts (winter). Himalayas; south to Cachar. Breeds *c.* 1200-2700 m.

9. (1749) WHITECOLLARED BLACKBIRD *Turdus albocinctus* R Myna+: 27 cm. Oak, chestnut, rhododendron forest often mixed with conifers. *Summer* (breeding): Himalayas, NE hill states; *c.* 2100-4000 m. *Winter*: Meghalaya, Cachar, Nagaland and Manipur.

10. (1757) Ssp. *kinnisii* of 1752. Sri Lanka.

11. (1753, 1754) Ssp. *nigropileus* of 1752. S Rajasthan, E Gujarat, east Vindhyas and Satpuras. W and E Ghats complex.

12. (1755) Ssp. *simillimus* of 1752. Nilgiris.

13. (1752) BLACKBIRD *Turdus merula* RM Myna+: 25-27 cm. Moist-deciduous and evergreen forest, sholas, wooded ravines, alpine meadows, etc. Baluchistan; Himàlayas; *c.* 3000-4500 m (breeding). Subcontinent: hills (summer), plains (winter).

14. (1761) FEA'S THRUSH *Turdus feai* M Myna±: 23 cm. Forested hills about *c.* 1500 m. Nepal, Nagaland, Manipur, Meghalaya and Cachar. Two records from Tamil Nadu from the Nilgiris and Pt.Calimere.

15. (1760) KESSLER'S THRUSH *Turdus kessleri* V Myna+: 27 cm. Low scrub and cultivated fields. Two records: (1) a party of four at Changu, Sikkim; (2) three sightings in Nepal between 3446 and 4328 m.

16. (1758) GREYHEADED THRUSH *Turdus rubrocanus* RM Myna+: 27 cm. Fir and horse-chestnut forest, open wooded country orchards, etc. W Himalayas. *Winter*: Assam and Meghalaya.

17. (1759) Eastern ssp. *gouldii* of 1758. Occasional winter visitor to Khasi and Cachar hills. Has strayed to Nepal and Uttar Pradesh.

PLATE 86

PLATE 86

1. (1731) PIED GROUND THRUSH *Zoothera wardii* RM Myna±: 22 cm. Evergreen biotope: well-wooded ravines, open country and edges of forest. *Summer*: Himalayas; NE hill states. *Winter*: E and W Ghats complex (passage?); Karnataka, Tamil Nadu, Sri Lanka. Seasonally between *c.* 750 and 2400 m.

2. (1737) SPOTTEDWINGED GROUND THRUSH *Zoothera spiloptera* R (endemic) Myna-: 21 cm. Forest, cardamom jungle, well-wooded country. Sri Lanka.

3. (1732) SIBERIAN GROUND THRUSH *Zoothera sibirica* M Myna±: 22 cm. Heavy Forest. Manipur hills; single record from Andaman Is.

4. (1739) PLAINBACKED MOUNTAIN THRUSH *Zoothera mollissima* R Myna+: 27 cm. Open grassy hillsides and boulder-strewn slopes with dense bushes and grass. Himalayas; NE hill states. Breeding *c.* 3000-4300 m.

5. (1746) LESSER BROWN THRUSH *Zoothera marginata* R Myna+: 25 cm. Dense undergrowth along watercourses in damp evergreen forest. Lower E Himalayas; NE hill states; Bangladesh.

6. (1740) LONGTAILED MOUNTAIN THRUSH *Zoothera dixoni* R Myna+: 27 cm. Dense forest, scrub and open bush country. Himalayas; NE hill states. Breeding *c.* 2100-4200 m.

7. (1745) LARGE BROWN THRUSH *Zoothera monticola* R Myna+: 28 cm. Small mountain streams in dense evergreen forest, and dense undergrowth. Himalayas; NE hill states; Bangladesh (Chittagong hill tracts). Seasonally from foothills to *c.* 3000 m.

8. (1741) GOLDEN or SMALLBILLED MOUNTAIN THRUSH *Zoothera dauma* R Myna+: 26 cm. Dense forest, edges of pasture-land, well-wooded margins of streams, bamboo brakes, etc. Himalayas; NE hill states; Bangladesh; also the southern W Ghats complex; Sri Lanka. Breeding *c.* 2100-3600 m.

9. (1765) DUSKY THRUSH *Turdus naumanni* M Myna±: 23 cm. Open fields, grassland and thinly wooded country. E Himalayas; NE hill states; *c.* 900-3000 m. Three records from Pakistan.

10. (1747) BLACKBREASTED THRUSH *Turdus dissimilis* R Myna±: 22 cm. Damp evergreen woods with ample undergrowth, and mixed oak and rhododendron forest; also scrub jungle. NE hill states; Bangladesh; *c.* 1200-2400 m.

11. (1762) DARK THRUSH *Turdus obscurus* M Myna±: 23 cm. Open forest. Arunachal Pradesh; NE hill states; Bangladesh. Also Tamil Nadu and Andamans.

12. (1764) REDTHROATED THRUSH *Turdus ruficollis ruficollis* M Myna+: 25 cm. Grassy slopes, forest edges, sparsely scrubbed fallow land, stubble fields and pastures, etc. Baluchistan; Himalayas; NE hill states; Bangladesh; up to *c.* 3000 m.

13. (1763) BLACKTHROATED THRUSH *Turdus ruficollis atrogularis*

14. (1767) REDWING *Turdus iliacus* V (M?) Myna±: 22 cm. Open forest and fields. Stray isolated records from NWFP(?) and Kashmir.

15. (1766) FIELDFARE *Turdus pilaris* V Myna+: 27 cm. Fields and orchards. Single record, Saharanpur, U.P.

16. (1768) MISTLE THRUSH *Turdus viscivorus* R Myna+: 28 cm. Mixed open forest of conifers, oak or birch; also open grassy hillsides and hill cultivation (winter). *Summer* (breeding): N Baluchistan, NWFP, Chitral, Gilgit east to Nepal, 1800-3900 m. *Winter*: Foothills and adjacent plains.

PLATE 87

1. (1474) PALEFOOTED BUSH WARBLER *Cettia pallidipes* R Sparrow-: 10 cm. Grass-and-bush or secondary jungle, and glades in evergreen or pine forest. Himalayan foothills from Dehra Dun east; NE hill states; E Ghats; Andaman Is. Bangladesh (winter). Seasonally, foothills to *c.* 1500 m.

2. (1476) JAPANESE BUSH WARBLER *Cettia diphone* V Sparrow±: Male 16 cm: Female 14 cm. Two records, Assam.

3. (1478) STRONGFOOTED BUSH WARBLER *Cettia montana* R Sparrow-: 11 cm. Bush-covered hillsides, bushes and undergrowth in open mixed forest. Himalayas from Swat east; NE hill states; Bangladesh (Chittagong region). Seasonally, foothills to *c.* 3300 m.

4. (1481) ABERRANT BUSH WARBLER *Cettia flavolivacea* R Sparrow-: 13 cm. Long grass, dense bushes and undergrowth in forest. Himalayas from Garhwal east; NE hill states. Seasonally, 700-3600 m.

5. (1486) RUFOUSCAPPED BUSH WARBLER *Cettia brunnifrons* R Sparrow-: 10 cm. Stunted juniper and furze bushes on stony hillsides and in open coniferous forest. Himalayas, *c.* 2700-4000 m (summer); also in the foothills and flood plains of the Brahmaputra, to Meghalaya (winter).

6. (1479) LARGE BUSH WARBLER *Cettia major* R Sparrow-: 13 cm. Dense rhododendron jungle in subalpine silver fir forest, or dwarf rhododendron thickets. Himalayas from Garhwal and Kumaon east; NE hill states. Seasonally, terai to 4000 m.

7. (1484) HUME'S BUSH WARBLER *Cettia acanthizoides* R Sparrow-: 11 cm. Dense ringal bamboo stands or open forest. Himalayas from Garhwal east; *c.* 2100-3700 m (summer).

8. (1493) BROWN BUSH WARBLER *Bradypterus luteoventris* R Sparrow-: 13 cm. Grassy downs, high grass and bracken-covered hillsides. E Himalayas; NE hill states. Seasonally, foothills to 3300 m.

9. (1494) PALLISER'S WARBLER *Bradypterus palliseri* R (endemic) Sparrow-: 14 cm. Undergrowth and dwarf bamboo in damp forest. Sri Lanka, above *c.* 900 m.

10. (1490) SPOTTED BUSH WARBLER *Bradypterus thoracicus* R Sparrow-: 13 cm. Bracken, dwarf juniper and other low scrub. Himalayas from Ladakh, Garhwal and Kumaon east. Seasonally, terai and plains to 4000+ m.

11. (1488) CETTI'S WARBLER *Cettia cetti* M Sparrow-: 12 cm. Partially submerged bulrushes, reeds, high grass and tamarisks. From Peshawar valley through the flood plains of the Indus; Sind; Punjab (Bahawalpur); and possibly Baluchistan; east to Harike Lake (Punjab) and Rajasthan.

12. (1555) BLACKBROWED REED WARBLER *Acrocephalus bistrigiceps* M Sparrow-: 13 cm. Dense cover near marshes, high grass and ricefields. Nepal; Assam, Manipur and lower Bengal. Bangladesh (?).

13. (1491) LARGEBILLED BUSH WARBLER *Bradypterus major* R Sparrow-: 13 cm. Low thorny scrub interspersed with rank grass and bracken fringing forest. W Himalayas from Gilgit and the Kagan valley east through Ladakh and Kashmir. Seasonally, *c.* 1200-3600 m.

14. (1549) THICKBILLED WARBLER *Acrocephalus aedon* M Bulbul±: 20 cm. Reeds and bushes in marshy places, tall grass and weeds in abandoned clearings, tea and coffee plantations. NE and peninsular India; also Gujarat. Rajasthan; Bangladesh; Andaman and Nicobar Is.

15. (1492) CHINESE BUSH WARBLER *Bradypterus tacsanowskius* M? Sparrow-: 14 cm. Grass and bushes, stubble fields, standing rice and reed-beds. Recorded from Jalpaiguri duars and Nepal terai (winter).

16. (1556) BLYTH'S REED WARBLER *Acrocephalus dumetorum* RM Sparrow-: 14 cm. Bushes, hedges, orchards, bamboo clumps and grain fields, mainly in deciduous biotope. *Breeding*: N Baluchistan. *Winter*: Subcontinent.

17. (1557) PADDYFIELD WARBLER *Acrocephalus agricola* RM Sparrow-: 13 cm. Reed-beds, elephant grass, sugarcane and wet rice cultivation. *Breeding*: N Baluchistan. *Winter*: Subcontinent. Absent in Sri Lanka.

18. (1555a) REED WARBLER *Acrocephalus scirpaceus* R Sparrow-: 13 cm. Reed-beds. Probably breeds in N Baluchistan; may winter on the Makran coast.

19. (1550) INDIAN GREAT REED WARBLER *Acrocephalus stentoreus* R Bulbul±: 19 cm. Reed-beds and shrubs around lakes, jheels, ponds, canals, etc. Subcontinent; Sri Lanka; Andaman Is. Breeding abundantly in Kashmir valley, *c.* 1800 m. Also in Bengal Salt Lakes; Kerala.

PLATE 87
1
2 ♀
2 ♂
3
4
5
6
7
8
9
10
11
12
13
14
15
16
17
18
19
19
John H. Dick

PLATE 88
1 imm
1 ♂
2
♀
3 ♂
6
7
4
5
♀
♀
11 ♂
♀
10 ♂
9 ♂
W
8 ♂
♀
13 S
15 S
12 ♂
14 S
16 W
17 W
15 W
16 S
17 S
John H. Dick

PLATE 88

1. (1564a) BARRED WARBLER *Sylvia nisoria* V Sparrow±: 16 cm. Bushes. Two records: Gilgit and Ladakh.

2. (1566) WHITETHROAT *Sylvia communis* M Sparrow-: 14 cm. Bushes, hedges amidst cultivation and scrub in semi-desert. *Summer*: C Baluchistan; Ladakh. *Autumn* (passage): Pakistan and NW India south to W Gujarat; Uttar Pradesh (Delhi, Kanpur). *Spring*: Makran coast.

3. (1565) ORPHEAN WARBLER *Sylvia hortensis* M Sparrow±: 15 cm. Shrubs on stony slopes and semi-desert. *Summer*: Baluchistan and NWFP. *Winter*: Pakistan; India from the Indus east, and south to Bihar, Karnataka and Tamil Nadu.

4. (1569) *Sylvia curruca minula*
5. (1570) *Sylvia curruca althaea*
6. (1567) *Sylvia curruca blythi*

LESSER WHITETHROAT *Sylvia curruca* M Sparrow-: 12 cm. Scrub jungle and bushes in semi-desert stony country. *Breeding (althaea)*: N Baluchistan; NWFP to Ladakh and Kashmir. *Winter*: (all sspp.): Semi-desert and drier areas of the Subcontinent.

7. (1571) DESERT WARBLER *Sylvia nana* M Sparrow-: 11 cm. Scrub in semi-desert, stony hillsides, and low vegetation on salty mudflats. Pakistan and NW India east to Haryana, Rajasthan and Kutch.

8. (1633) STOLICZKA'S TIT-WARBLER *Leptopoecile sophiae* R Sparrow-: 10 cm. Dwarf juniper and other scrub and thickets of *Hippophae* and willows. Baltistan, Gilgit, Ladakh. Seasonally, *c.* 1800-3900 m.

9. (1634) Ssp. *obscura* of 1633. From NW Nepal east to Arunachal, *c.* 2800-4000 m.

10. (1629) GOLDCREST *Regulus regulus* R Sparrow-: 8 cm. Conifer forest. Himalayas, seasonally from *c.* 1500 m to upper limit of conifer forest.

11. (1571a) MENETRIES'S WARBLER *Sylvia mystacaea* M Sparrow-: 12 cm. Bushes of willow and tamarisk. Rare and localised breeding visitor to W C Baluchistan.

12. (1632) CRESTED TIT-WARBLER *Leptopoecile elegans* R Sparrow-: 10 cm. Fir forest and juniper scrub. Presumably occurring in Arunachal Pradesh adjacent to SE Tibet where resident. Seasonally, *c.* 2800-4300 m.

13. (1517) ASHY WREN-WARBLER *Prinia socialis* R Sparrow-: 13 cm. *Sarpat* grass, reeds, scrub and grass around cultivation, bracken-covered slopes, gardens. Peninsular India south of Narbada river and S Bihar; Sri Lanka.

14. (1515) Ssp. *stewarti* of 1517. Upper Indus system (Pakistan) and India from Himalayan foothills south to Narbada river, east to Orissa, N Andhra and NE India.

15. (1507) HODGSON'S WREN-WARBLER *Prinia cinereocapilla* R Sparrow-: 11 cm. Dense jungle, secondary growth and trees. Himalayan foothills, duns and bhabar from Kumaon east.

16. (1501) RUFOUS WREN-WARBLER *Prinia rufescens* R Sparrow-: 11 cm. Stands of long grass, secondary growth and weeds on open forested hillsides. E Himalayas; NE hill states; Bangladesh; Orissa; E Ghats (Vishakhapatnam Hills). Foothills to *c.* 1800 m.

17. (1503) FRANKLIN'S WREN-WARBLER *Prinia hodgsonii* R Sparrow-: 11 cm. Scrub and grass jungle, undergrowth in deciduous forest, mangrove swamps, reed-beds, etc. Subcontinent. Absent in Andaman and Nicobar Is.

PLATE 89

1. (1543) PALLAS'S GRASSHOPPER WARBLER *Locustella certhiola* M Sparrow-: 13 cm. Swamps, reed-beds and ricefields. NE and C India; Bangladesh; Sri Lanka; Andaman and Nicobar Is.

2. (1546) BROADTAILED GRASS WARBLER *Schoenicola platyura* R Sparrow+: 18 cm. Tall grass and scrub covered hillsides, reeds and grass in marshy depressions. Southern W Ghats complex from Belgaum south; Sri Lanka.

3. (1545) GRASSHOPPER WARBLER *Locustella naevia* M Sparrow-: 13 cm. Edges of reservoirs and swampy depressions, damp grass and tamarisk jungle. Subcontinent.

4. (EL) Ssp. *obscurior* of 1545. N Caucasus. SW Iran (winter).

5. (1548) STRIATED MARSH WARBLER *Megalurus palustris* R Bulbul+: 25 cm. Grass and reeds in marshes, tall grass and scrub in cultivation. N Pakistan; N and NE India; Bangladesh.

6. (1547) BRISTLED GRASS WARBLER *Chaetornis striatus* R Sparrow+: 20 cm. Coarse grassland intermixed with bushes and standing ricefields. Subcontinent. Absent in Sri Lanka; Andaman and Nicobar Is.

7. (1508) STREAKED WREN-WARBLER *Prinia gracilis* R Sparrow-: 13 cm. Tamarisk, grass and scrub jungle in sandy, semi-desert country. Baluchistan; NW Pakistan; NW India, Himalayan foothills, and Gangetic plain east to Brahmaputra river. Bangladesh.

8. (1506) RUFOUSFRONTED WREN-WARBLER *Prinia buchanani* R Sparrow-: 12 cm. Arid scrub and grass jungle in semi-desert. Pakistan and N and C India.

9. (1544) STREAKED GRASSHOPPER WARBLER *Locustella lanceolata* M Sparrow-: 13 cm. Dense bush and grassland, sugarcane, rice stubbles. N and NE India; Bangladesh; Andaman and Nicobar Is.

10. (1495) MOUSTACHED SEDGE WARBLER *Acrocephalus metanopogon* R Sparrow±: 15 cm. Reed-beds and partially submerged vegetation in jheels, etc. *Breeding*: Baluchistan; Punjab; Kumaon terai. *Winter*: N and NW India.

11. (1529) BLACKTHROATED HILL WARBLER *Prinia atrogularis* R Sparrow-, with long tail: overall 17 cm. Open scrub and grass jungle on hillsides. E Himalayas; NE hill states; Bangladesh; *c.* 900-2500 m.

12. (1511) PLAIN WREN-WARBLER *Prinia subflava* R Sparrow-: 13 cm. Mixed thorn scrub and tall grass jungle, secondary growth, cultivation, etc. Subcontinent. Absent in Andaman and Nicobar Is.

13. (1521) JUNGLE WREN-WARBLER *Prinia sylvatica* R Sparrow±: 15 cm. Low bush jungle mixed with coarse grass. Subcontinent from Kangra east to NW Bangladesh. Sri Lanka. Absent in Andaman and Nicobar Is.

14. (1524) Ssp. *sindiana* of 1525 Pakistan; Punjab, Haryana.

15. (1525) YELLOWBELLIED WREN-WARBLER *Prinia flaviventris* R Sparrow-: 13 cm. Humid grassland mixed with bushes, elephant grass, reeds and thin secondary growth. Pakistan; N and NE India; Bangladesh.

16. (1532) Disjunct Eastern ssp. *cinerascens* of 1531. R Sparrow-, with long tail: overall 16 cm. Elephant grass near large rivers and in swamps. Assam plains of the Brahmaputra river system; Bangladesh.

17. (1531) LONGTAILED GRASS WARBLER *Prinia burnesii* R Sparrow-, with long tail: overall 17 cm. Stretches of *sarkhan* grass and bushes, near rivers. Pakistan: flood plains of the Indus river system.

18. (1533) STREAKED SCRUB WARBLER *Scotocerca inquieta* R Sparrow-: 10 cm. Scrub and grass clumps on arid, rocky hillsides. Baluchistan, NW Pakistan; foothills to *c.* 3000 m.

19. (1527) BROWN HILL WARBLER *Prinia criniger* R Sparrow±: 16 cm. Grass and scrub jungle and open pine forest on hillsides. Baluchistan, N Pakistan and Himalayan foothills; NE hill states; Bangladesh. Seasonally, *c.* 600-3100 m.

20. (1498) STREAKED FANTAIL WARBLER *Cisticola juncidis* R Sparrow-: 10 cm. Tall grass, reed-beds, ricefields, dry grassland, etc. Subcontinent. Absent in Andaman Is.

21. (1534) LARGE GRASS WARBLER *Graminicola bengalensis* R Sparrow±: 16 cm. Tall grass and reeds. From W Nepal terai east through Bengal and flood plains of the Brahmaputra; Bangladesh; Manipur.

22. (1496) FANTAIL WARBLER *Cisticola exilis* R Sparrow-: 10 cm. Tall coarse grass and bracken scrub on open hillsides and lowlands. Southern W Ghats complex; Sub-Himalayan terai; NE India; Bangladesh.

1
2
2
3
4
5
6
7
8
9
10
11 W
11 S
12
13
14
15
16
17
18
19
20
21
22 ♂ br +
22 ♀
John H. Dick

PLATE 90

PLATE 90

1. (1564) UPCHER'S WARBLER *Hippolais languida* M Sparrow-: 14 cm. Stunted bush-covered stony hill slopes. Baluchistan (Quetta).

2. (1577) PLAIN LEAF WARBLER *Phylloscopus neglectus* M Sparrow-: 10 cm. Low bushes, juniper forest, tamarisks and acacias. *Summer*: N Baluchistan; NWFP east to Kashmir. *Winter*: W Himalayan foothills south through Sind and S Baluchistan. Seasonally, foothills to *c.* 3000 m.

3. (1562) BOOTED WARBLER *Hippolais caligata* RM Sparrow-: 12 cm. Deciduous scrub jungle. *Breeding*: Throughout Baluchistan; NWFP; Sind. *Winter*: Subcontinent; Sri Lanka.

4. (1574) BROWN LEAF WARBLER, or CHIFFCHAFF *Phylloscopus collybita* M Sparrow-: 10 cm. Bushes, hedges, gardens, scrub jungle. *Summer*: Baltistan, Gilgit, Ladakh, Rupshu, Lahul, Spiti, *c.* 2500-4400 m. *Winter*: Himalayan foothills and Subcontinent.

5. (1578) TYTLER'S LEAF WARBLER *Phylloscopus tytleri* M Sparrow-: 10 cm. Coniferous forest; also dwarf willows and birches. *Summer*: Extreme NW Himalayas to Kashmir; *c.* 2400 m to timber line. *Winter*: Subcontinent south to Kerala.

6. (1579) TICKELL'S LEAF WARBLER *Phylloscopus affinis* M Sparrow-: 10 cm. Scrub, secondary jungle, well-wooded country. *Summer*: Himalayas, *c.* 2700 m. to timber line. *Winter*: Subcontinent.

7. (1580) Buffbellied ssp. *arcanus* of 1579. Kathmandu valley and W Nepal terai (winter).[1]

8. (1581) OLIVACEOUS LEAF WARBLER *Phylloscopus griseolus* RM Sparrow-: 10 cm. Deciduous biotope: stony hillsides with sparse bushes. *Summer*: N Baluchistan; NW Himalayas; *c.* 2400-4500 m. *Winter*: Subcontinent.

9. (1593) BROOKS'S LEAF WARBLER *Phylloscopus subviridis* R Sparrow-: 10 cm. Coniferous forest, bushes, olive groves and acacias. NWFP; NW Himalayas; *c.* 2100-3600 m. *Winter*: N India.

10. (1592) YELLOWBROWED LEAF WARBLER *Phylloscopus inornatus* RM Sparrow-: 10 cm. Scrub, gardens, groves and open forest. *Summer*: W Himalayas up to *c.* 3600 m. *Winter*: Subcontinent.

11. (1586) DUSKY LEAF WARBLER *Phylloscopus fuscatus* M Sparrow-: 10 cm. Scrub jungle, low bushes, reeds and long grass around pools, hedges and standing crops. E Himalayas; N and NE India; Bangladesh; Andaman Is. Up to *c.* 1400 m.

12. (1582) SMOKY WILLOW WARBLER *Phylloscopus fuligiventer* R Sparrow-: 10 cm. Low scrub, boulder-strewn alpine meadows, along banks of watercourses, etc. E Himalayas. *Winter*: W Nepal terai and duns east. Seasonally, foothills to *c.* 4500 m.

13. (1612a) RADDE'S LEAF WARBLER *Phylloscopus schwarzi* V(?). Two sight records from E & C Nepal.

14. (1588) ORANGEBARRED LEAF WARBLER *Phylloscopus pulcher* R Sparrow-: 10 cm. Subalpine mixed conifer-rhododendron and birch forest, scrub. Himalayas; NE hill states. Seasonally, *c.* 500-4300 m.

15. (1594) PALLAS'S LEAF WARBLER *Phylloscopus proregulus* R Sparrow-: 9 cm. Conifer or mixed forest, woods and bush-covered hillsides. Himalayas, *c.* 2200-4200 m. *Winter*: Foothills; NE India; Bangladesh.

16. (1599) GREYFACED LEAF WARBLER *Phylloscopus maculipennis* RM Sparrow-: 9 cm. Open mixed forest of oak and rhododendron; also mixed deciduous forest with plenty of undergrowth. Himalayas from Kashmir to Arunachal, Nagaland and NE hill states. Seasonally from foothills to 3500 m.

17. (1612) BLACKBROWED LEAF WARBLER *Phylloscopus cantator* R Sparrow-: 9 cm. Dense evergreen, open deciduous and mixed forest. *Breeding*: Assam and ?Manipur. *Winter*: NE India including Bangladesh (Chittagong region). Also Sikkim, Bhutan and E Nepal.

18. (1606) LARGE CROWNED LEAF WARBLER *Phylloscopus occipitalis* M Sparrow-: 10 cm. Dry- and moist-deciduous, evergreen or subtropical wet forest. *Summer*: W Himalayas from NWFP and Baltistan to Garhwal and Kumaon *c.* 1800-3200 m. *Winter*: Subcontinent.

19. (1607) Eastern ssp. *coronatus* of 1606. *Winter*: Sikkim; NE India; Bangladesh.

20. (1609) BLYTH'S LEAF WARBLER *Phylloscopus reguloides* R Sparrow-: 9 cm. Oak, rhododendron, conifer, evergreen wet or pine forest. Himalayas; NE hill states; Bangladesh. Seasonally, plains and foothills to *c.* 3600 m.

21. (1605) GREENISH LEAF WARBLER *Phylloscopus trochiloides* M Sparrow-: 10 cm. Gardens, wooded compounds, groves and open deciduous forest. *Summer*: Himalayas; NE hill states; *c.* 2700-4200 m. *Winter*: Subcontinent; Andamans; Sri Lanka.

22. (1601) LARGEBILLED LEAF WARBLER *Phylloscopus magnirostris* M Sparrow-: 12 cm. Deciduous or evergreen forest. *Summer*: Himalayas, *c.* 1800-3600 m. *Winter*: NE and peninsular India; Sri Lanka; Andaman Is.

[1] Synonymised with nominate *Cettia flavolivacea* of plate 87, Hence omit.

PLATE 91

1. (1450) YELLOWBELLIED FANTAIL FLYCATCHER *Rhipidura hypoxantha* R Sparrow-: 8 cm. Evergreen biotope: mixed coniferous and birch or rhododendron forest. Himalayas; NE hill states; Bangladesh (Chittagong region). Seasonally, foothills to tree-line.

2. (1615) BLACKBROWED FLYCATCHER-WARBLER *Seicercus burkii* R Sparrow-: 10 cm. Evergreen, dense deciduous or mixed forest. *Breeding*: Himalayas; NE hill states; Bangladesh. *Winter*: foothills and plains of N India south to Maharashtra and NE Andhra. Seasonally, foothills to *c.* 3700 m.

3. (1626) WHITETHROATED FLYCATCHER-WARBLER *Abroscopus albogularis* R Sparrow-: 8 cm. Bamboo jungle, secondary scrub and moist-deciduous forest. E Himalayas; NE hill states; Bangladesh (Chittagong region) *c.* 300-1800 m.

4. (1624) BLACKFACED FLYCATCHER-WARBLER *Abroscopus schisticeps* R Sparrow-: 9 cm. In forest with undergrowth of scrub or bamboo, near streams. Himalayas from Garhwal east; NE hill states; *c.* 1500-2500 m.

5. (1613) ALLIED FLYCATCHER WARBLER *Seicercus affinis* R Sparrow-: 10 cm. Dense evergreen, broadleafed or pine forest. E Himalayas; NE hill states. Seasonally, foothills to *c.* 2300 m.

6. (1622) YELLOWBELLIED FLYCATCHER-WARBLER *Abroscopus superciliaris* R Sparrow-: 9 cm. Mixed bamboo forest, secondary and scrub jungle, and near streams. E Himalayas; NE hill states; Bangladesh (Chittagong region). Seasonally, foothills to *c.* 2400 m.

7. (1617) GREYHEADED FLYCATCHER-WARBLER *Seicercus xanthoschistos* R Sparrow-: 10 cm. Coniferous or open evergreen forest, scrub jungle and gardens. Himalayas; NE hill states. Seasonally, foothills to *c.* 2700 m.

8. (1449) GREYHEADED FLYCATCHER *Culicicapa ceylonensis* R Sparrow-: 9 cm. Deciduous or evergreen forest, sholas, secondary and mixed bamboo forest. *Summer*: Himalayas (up to *c.* 3000 m): NE hill states; Bangladesh; C. India; E and W Ghats; Sri Lanka. *Winter*: Subcontinent.

9. (1620) GREYCHEEKED FLYCATCHER-WARBLER *Seicercus poliogenys* R Sparrow-: 10 cm. Open evergreen forest and dense bamboo jungle. Himalayas from Uttar Pradesh; NE hill states; Bangladesh (Chittagong region). Seasonally, foothills to *c.* 3000 m.

10. (1627) BROADBILLED FLYCATCHER-WARBLER *Abroscopus hodgsoni* R Sparrow-: 10 cm. Dense scrub and bamboo along forest edges. E Himalayas; NE hill states. Breeds between *c.* 1100 and 2700 m.

11. (1538) TAILOR BIRD *Orthotomus sutorius* R Sparrow-: 13 cm. Scrub country near cultivation, gardens, wooded compounds and deciduous jungle. Subcontinent. Absent in Andaman and Nicobar Is.

12. (1541) GOLDENHEADED TAILOR BIRD *Orthotomus cucullatus* R Sparrow-: 12 cm. Evergreen biotope: scrub, grass and bamboo jungle, secondary growth. E Himalayas; NE hill states; up to *c.* 1800 m.

13. (1621) CHESTNUT-HEADED FLYCATCHER-WARBLER *Seicercus castaniceps* R Sparrow-: 10 cm. Dense subtropical, pine and wet-temperate forest. E Himalayas; NE hill states; Bangladesh (Chittagong region).

14. (1540) BLACKNECKED TAILOR BIRD *Orthotomus atrogularis* R Sparrow-: 13 cm. Heavy scrub and edges of evergreen forest. E Himalayas; NE hill states; up to *c.* 1800 m.

15. (1471) DULL SLATYBELLIED GROUND WARBLER *Tesia cyaniventer* R Sparrow-: 9 cm. By small streams in dense shady evergreen and moist-deciduous forest. Himalayas from Garhwal east; NE hill states; Bangladesh (Chittagong region). Seasonally, foothills to *c.* 2550 m.

16. (1472) SLATYBELLIED GROUND WARBLER *Tesia olivea* R Sparrow-: 9 cm. Dense undergrowth of ferns, nettles and weeds in humid tropical forest. E Himalayas; NE hill states.

17. (1473) CHESTNUT-HEADED GROUND WARBLER *Tesia castaneocoronata* R Sparrow-: 8 cm. Dense undergrowth of bushes, ferns, nettles or bamboo, near streams in high forest. Himalayas; NE hill states; Bangladesh (Chittagong region). Seasonally, foothills to *c.* 3900 m.

1
2
3
4
5
6
7
8
9
10
11 ♂ br +
11 ♂ br –
12
13
14 ♂
♀
15
16
17
John H. Dick

PLATE 92
1
3 ♂
4 ♂
2 ♂
2 ♂
5
2 ♀
6
7
8
♀
10 ♂
11
9
♀
12 ♂
13
14 ♀
14 ♂
♀
15 ♂
17 ♂
18 ♂
♀
16 ♂
18 ♀
17 ♀
20 ♂
21 ♂
♀
19 ♂
John H. Dick

PLATE 92

1. (1470) GREY THICKHEAD or MANGROVE WHISTLER *Pachycephala grisola* R Sparrow+: 17 cm. Mangroves and shrubby trees. Sundarbans of Bengal and Bangladesh; Andaman Is.

2. (1411) REDBREASTED FLYCATCHER *Muscicapa parva* M Sparrow-: 13 cm. Groves, forest plantations, gardens with large trees, etc. Pakistan and India east to Bengal, south to Karnataka.

3. (1412) Eastern ssp. *albicilla* of 1411. E and NE India, south though the peninsula to Kerala.

4. (1413) KASHMIR REDBREASTED FLYCATCHER *Muscicapa subrubra* R Sparrow-: 13 cm. Temperate mixed forest, gardens, tea estates, etc. *Summer*: Kashmir and Pir Panjal range, *c.* 1800-2700 m. *Winter*: Sri Lanka.

5. (1403) SPOTTED FLYCATCHER *Muscicapa striata* R Sparrow-: 13 cm. Open forest, especially pine. *Summer*: Baluchistan; NW Himalayas. *Breeding* between *c.* 2100 and 3300 m. *Autumn* (passage): Pakistan and NW India.

6. (1408) BROWNBREASTED FLYCATCHER *Muscicapa muttui* R Sparrow-: 13 cm. Dense evergreen forest and jungle. *Summer*: NE hill states above 1200 m. *Winter*: SW India; Sri Lanka. Bangladesh (passage).

7. (1406) SOOTY FLYCATCHER *Muscicapa sibirica* R Sparrow-: 13 cm. Open forest of conifer and oak. Himalayas; NE hill states. Seasonally, foothills to timber-line.

8. (1409) RUFOUSTAILED FLYCATCHER *Muscicapa ruficauda* M Sparrow±: 14 cm. Conifer, deciduous and evergreen forest. *Summer*: W Himalayas, *c.* 2100-3600 m. *Winter*: SW India. Kutch, Bharatpur and E Ghats (passage).

9. (1410) FERRUGINOUS FLYCATCHER *Muscicapa ferruginea* R Sparrow-: 10 cm. Wet forest of fir and oak, and dense mixed jungle. E Himalayas; NE hill states; *c.* 1800-3300 m.

10. (1427) BLACK-AND-ORANGE FLYCATCHER *Muscicapa nigrorufa* R Sparrow-: 13 cm. Dense evergreen sholas with ample undergrowth, edges of coffee plantations, and dank ravines. W Ghats complex from Wynaad south.

11. (1407) BROWN FLYCATCHER *Muscicapa latirostris* RM Sparrow±: 14 cm. Woodland, scrub, cultivation, etc. Himalayan foothills; India (hilly regions); W Ghats complex from N Kanara south; E Ghats (?). *Winter*: Subcontinent; Sri Lanka; Andamans and Nicobars.

12. (1414) ORANGEGORGETED FLYCATCHER *Muscicapa strophiata* R Sparrow-: 13 cm. Oak, rhododendron, conifer, birch and shady mixed forest. Himalayas; NE hill states. *Winter*: South of the Brahmaputra to Bangladesh (Chittagong). Seasonally, foothills to *c.* 3700 m.

13. (1402) OLIVE FLYCATCHER *Rhinomyias brunneata* M Sparrow±: 14 cm. Forest, sometimes gardens Great and Little Nicobar Is. Possibly Andamans.

14. (1419, 1420) LITTLE PIED FLYCATCHER *Muscicapa westermanni* R Sparrow-: 10 cm. Dense evergreen or deciduous forest, open woodland, orchards, etc. *Summer*: Himalayas from Uttar Pradesh; NE hill states. *Winter*: foothills and plains, C and E India; Bangladesh. Seasonally, foothills to *c.* 2700 m.

15. (1421) WHITEBROWED BLUE FLYCATCHER *Muscicapa superciliaris* R Sparrow-: 10 cm. Open deciduous forest, groves, gardens and orchards, up to *c.* 3200 m. *Summer*: Himalayas; NE hill states. *Winter*: N and C India.

16. (1429) SMALL NILTAVA *Muscicapa macgrigoriae* R Sparrow-: 11 cm. Bushes near streams, shady glades and mixed heavy secondary jungle. Himalayas from Kumaon east NE hill states. Seasonally, foothills to *c.* 2100 m.

17. (1424) Eastern ssp. *minuta* of 1423. E Himalayas; NE hill states.

18. (1423) SLATY BLUE FLYCATCHER *Muscicapa leucomelanura* R Sparrow-: 10 cm. Dense mixed forest of oak and rhododendron, and woodland with ample undergrowth. W Himalayas, seasonally, foothills to *c.* 4000 m.

19. (1428) LARGE NILTAVA *Muscicapa grandis* R Bulbul±: 20 cm. Dense humid forest and secondary jungle on steep hillsides and in ravines. E Himalayas; NE hill states. Seasonally, foothills to *c.* 2700 m.

20. (1416) Ssp. *leucops* of 1415. NE hill states; Bangladesh (Chittagong region).

21. (1415) WHITEGORGETED FLYCATCHER *Muscicapa monileger* R Sparrow-: 11 cm. Dense bush jungle, scrubby ravines or thick undergrowth in tropical forest. E Himalayas, foothills to *c.* 2000 m.

PLATE 93

1. (1442) TICKELL'S BLUE FLYCATCHER *Muscicapa tickelliae* R Sparrow-: 14 cm. Moist-deciduous or evergreen biotope: scrub and bamboo jungle, groves and wooded gardens. N and peninsular India, excepting NW arid parts; Bangladesh; Sri Lanka.

2. (1426) SAPPHIREHEADED FLYCATCHER *Muscicapa sapphira* R Sparrow-: 12 cm. Evergreen forest. E Himalayas; NE hill states. Seasonally, 800-2600 m.

3. (1447) PIGMY BLUE FLYCATCHER *Muscicapella hodgsoni* R Sparrow-: 8 cm. Dense, tall forest and secondary scrub at edge of clearings or along hill streams. E Himalayas; NE hill states. Seasonally, foothills to *c.* 3500 m.

4. (1418) RUSTYBREASTED BLUE FLYCATCHER *Muscicapa hodgsonii* R Sparrow-: 13 cm. Conifer, oak, bamboo and rhododendron forest, and thick scrub. E Himalayas; NE hill states; Bangladesh. Seasonally, foothills to 3900 m.

5. (1417) RUFOUSBREASTED BLUE FLYCATCHER *Muscicapa hyperythra* R Sparrow-: 11 cm. Dense primary forest with luxuriant undergrowth. Himalayas from Kumaon east; NE hill states. Seasonally 300-3000 m.

6. (1440) BLUETHROATED FLYCATCHER *Muscicapa rubeculoides* R Sparrow-: 14 cm. Moist-deciduous or evergreen biotope: forest with ample undergrowth, secondary or bamboo jungle. *Summer*: Himalayan foothills; NE hill states; Bangladesh. *Winter*: continental and peninsular India; Sri Lanka. Seasonally, foothills to *c.* 2100 m.

7. (1441) LARGEBILLED BLUE FLYCATCHER *Muscicapa banyumas* R Sparrow-: 14 cm. Shady ravines and undergrowth in dense humid forest. E Himalayas; NE hill states; Bangladesh. Seasonally, foothills to *c.* 2600 m. Rare and little known.

8. (1436) BROOKS'S FLYCATCHER *Muscicapa poliogenys* R Sparrow-: 14 cm. Tropical evergreen and moist-deciduous forest. E Himalayas; NE hill states (up to *c.* 1500 m); E Ghats complex in Orissa and Andhra Pradesh (up to *c.* 1000 m).

9. (1435) WHITEBELLIED BLUE FLYCATCHER *Muscicapa pallipes* R Sparrow±: 15 cm. Undergrowth in evergreen forest, sholas, *Strobilanthes* and cardamom ravines. W Ghats complex from *c.* 19°N south. Foothills to *c.* 1500 m.

10. (1445) VERDITER FLYCATCHER *Muscicapa thalassina* R Sparrow±: 15 cm. Deciduous, evergreen or light conifer forest. *Summer*: Himalayas; NE hill states; *c.* 1200-3000 m. *Winter*: Subcontinent, but not Sri Lanka.

11. (1439) PALE BLUE FLYCATCHER *Muscicapa unicolor* R Sparrow±: 16 cm. Moist-deciduous or evergreen biotope: secondary or bamboo jungle, dense humid forest. Himalayas from Garhwal east; NE hill states; Bangladesh (Chittagong hill tracts). Seasonally, foothills to *c.* 1800 m.

12. (1444) DUSKY BLUE FLYCATCHER *Muscicapa sordida* R (endemic) Sparrow-: 14 cm. Forest and well-wooded ravines. Sri Lanka.

13. (1446) NILGIRI VERDITER FLYCATCHER *Muscicapa albicaudata* R Sparrow±: 15 cm. Sholas, forest glades, coffee and cardamom plantations, etc. Southern W Ghats complex.

14. (1432) RUFOUSBELLIED NILTAVA *Muscicapa sundara* R Sparrow±: 15 cm. Dense undergrowth in forest, secondary growth and brush-covered hillsides. *Summer*: Himalayas; NE hill states; *c.* 900-3200 m. *Winter*: foothills and adjacent plains; Bangladesh (Chittagong region).

15. (1434) WHITETAILED BLUE FLYCATCHER *Muscicapa concreta* R Bulbul-: 18 cm. Deep forest. Said to breed in the Patkai Range (Burma border) above 1500 m; obtained in winter in the Margherita-Dibrugarh area.

16. (1433) RUFOUSBELLIED BLUE FLYCATCHER *Muscicapa vivida* R Sparrow+: 18 cm. Dense brushwood in evergreen forest. Arunachal Pradesh, the eastern hills of Manipur; Nagaland(?). Seasonally, 1500-2700 m.

1♂ ♀
2♂ ♂ 2 ad- 2♀
♀ 3♂
♀ 4♂
♀ 5♂
6♂ ♀
7♂
8
♀ 9♂
♀ 10♂
♀ 11♂
12♂
♀ 13♂
♀ 14♂
♀ 15♂
♀ 16♂
John H. Dick

PLATE 94
2 ♂ ad –
♀
1 ♂
2 ♂
3
4
5
imm
♀
6 imm
9
7
6
7 imm
imm
♀
8
13 ♂
10
imm
12
11
16
15
14 imm
♀
17
14
19 ♂
18
John H. Dick

PLATE 94

1. (1465) BLACKNAPED FLYCATCHER *Hypothymis azurea* R Sparrow±: 16 cm. Evergreen and moist-deciduous biotope: forest and plantations, mixed bamboo jungle, etc. Himalayan foothills from Gorakhpur; continental and peninsular India; Andaman and Nicobar Is.; Bangladesh; Sri Lanka.

2. (1461) PARADISE FLYCATCHER *Terpsiphone paradisi* RM Bulbul±: 20 cm. Male (with tail streamers) overall 50 cm. Well-watered and shady forest, plantations, groves and gardens. Subcontinent. Also Andamans and Nicobars.

3. (1458) Whitespotted ssp. *albogularis* of 1455. Most of peninsular India south of the Vindhya Range.

4.(1451) WHITEBROWED FANTAIL FLYCATCHER *Rhipidura aureola* R Bulbul-: 17 cm. Dry open country, wooded compounds, parkland, scrub and bamboo jungle. Subcontinent excepting NE parts and Andaman and Nicobar Is.

5. (1455) WHITETHROATED FANTAIL FLYCATCHER *Rhipidura albicollis* R Bulbul-: 17 cm. Shady places in forest, well-wooded country, gardens, groves and secondary scrub. Himalayas; NE hill states (up to *c.* 2700 m); peninsular India.

6. (940) BAYBACKED SHRIKE *Lanius vittatus* R Bulbul-: 18 cm. Dry-deciduous open thorn scrub jungle, outskirts of cultivation. Subcontinent but not in Andamans, Nicobars and Sri Lanka.

7. (943) Pale brown ssp. *isabellinus* of 941. *Winter*: Pakistan and NW India east to Bihar, and south to E Madhya Pradesh, Gujarat and W Maharashtra. Recorded from Nepal.

8. (942) Rufous ssp. *phoenicuroides* of 941. *Summer*: N Baluchistan. *Autumn*: Passage through Pakistan, east to Haryana, C Rajasthan, Gujarat.

9. (949) BROWN SHRIKE *Lanius cristatus* M Bulbul±: 19 cm. Dry-deciduous and semi-evergreen scrub and scattered bushes. Locally Subcontinent. Also Lakshadweep, Andamans, Nicobars, Sri Lanka and Maldives.

10. (938) BURMESE SHRIKE *Lanius collurioides* R Bulbul+: 23 cm. Evergreen scrub and semi-cultivation. N Cachar; Nagaland(?); Manipur; Bangladesh(?).

11. (937) LESSER GREY SHRIKE *Lanius minor* V Myna±: 23 cm. Two records. Baluchistan (Quetta and Chaman). Open semi-desert and cultivation scattered with thorn bushes.

12. (933) GREY SHRIKE *Lanius excubitor* RM Myna+: 25 cm. Tropical thorn and dry-deciduous biotope. Pakistan from NWFP and Baluchistan east through continental and peninsular India and Bangladesh in appropriate facies.

13. (941) REDBACKED SHRIKE *Lanius collurio* M Bulbul-: 17 cm. Open scrub jungle and cultivation. Pakistan east to W Rajasthan and N Gujarat.

14. (951) WOODCHAT SHRIKE *Lanius senator* V? Bulbul-: 17 cm. An unconfirmed record from Baluchistan (Quetta).

15. (945) GREYBACKED or TIBETAN SHRIKE *Lanius tephronotus* R Bulbul+: 25 cm. Bushes, open scrub and cultivation. Himalayas from Baltistan, Ladakh and N Kashmir east. *Winter*: foothills and plains of N and NE India. Seasonally, foothills to *c.* 4500 m.

16. (1062) WAXWING *Bombycilla garrulus* V Bulbul-: 18 cm. Fruiting bushes and trees. Has occurred in Baluchistan (Quetta), NWFP, Kashmir and Nepal.

17. (946) RUFOUSBACKED SHRIKE *Lanius schach* R Bulbul+: 25 cm. Open wooded country and cultivation. Subcontinent. Absent in Andaman and Nicobar Is.

18. (948) Blackheaded ssp. *tricolor* of 946. N and NE India; Bangladesh.

19. (1063) GREY HYPOCOLIUS *Hypocolius ampelinus* V Bulbul+: 25 cm. Semi-desert and open deciduous scrub jungle. Baluchistan, Sind, Kutch and Maharashtra.

1
2
3
4
5
6 W
6 S
7 W
7 S
8
9
10 W
10 S
11
12
13 S
14 W
14 S
15
imm
16 S
17 S
imm
18
19
19 W
19S
W
20 S
20 imm
John H. Dick

PLATE 95

1. (1857) PADDYFIELD PIPIT Migrant ssp. *richardi* of 1858. Sparrow+: 17 cm. Cultivation, bare hillsides and sandy areas. E and NE India. Patchily throughout Subcontinent; also Andaman Is.

2. (1858) PADDYFIELD PIPIT *Anthus novaeseelandiae* R Sparrow±: 15 cm. Grassland, stubble fields, fallows and marshy ground. Subcontinent; Sri Lanka.

3. (1861) TAWNY PIPIT *Anthus campestris* M Sparrow±: 15 cm. Sparsely shrubbed semi-desert, fallow land, pastures, ploughed fields. Subcontinent.

4. (1868) BROWN ROCK PIPIT *Anthus similis* RM Bulbul±: 20 cm. Grassy slopes and plains, sparsely scrubbed country, fallow land, sand dunes, etc. Subcontinent.

5. (1856) MEADOW PIPIT *Anthus pratensis* M Sparrow±: 15 cm. Grassy meadows and pastureland. NWFP.

6. (1864) REDTHROATED PIPIT *Anthus cervinus* M Sparrow±: 15 cm. Grassy wet ground, and stubble fields. Himalayan foothills, continental and peninsular India; Andaman and Nicobar Is.

7. (1865) VINACEOUSBREASTED PIPIT *Anthus roseatus* RM Sparrow±: 15 cm. Alpine meadows and boulder-strewn grassy slopes, marshes, jheels and ricefields. *Summer*: Himalayas from Baltistan and Ladakh east; mostly above timber-line. *Winter*: N Pakistan foothills; continental India; Bangladesh; up to *c.* 1500 m.

8. (1852) INDIAN TREE PIPIT *Anthus hodgsoni* RM Sparrow±: 15 cm. Grassy slopes, rocky ground and glades in open forest. *Summer*: Subalpine Himalayas from *c.* 2700 to timber-line. *Winter*: Continental and peninsular India; Bangladesh; Sri Lanka.

9. (1854) TREE PIPIT *Anthus trivialis* M Sparrow±: 15 cm. Grassy slopes, cultivation, stubble fields, groves and light forest. *Summer*: NW Himalayas from *c.* 2700 m to timber-line. *Winter*: Most of continental and peninsular India from the Himalayan foothills south.

10. (1871) WATER PIPIT or ALPINE PIPIT *Anthus spinoletta* M Sparrow±: 15 cm. Damp ground, marshes, canals, irrigated ricefields, etc. Most of the Subcontinent.

11. (1870) NILGIRI PIPIT *Anthus nilghiriensis* R Sparrow+: 17 cm. Rolling downs and grassy hilltops, woodland, coffee estates. Southern W Ghats complex, above *c.* 1000 m.

12. (1873) UPLAND PIPIT *Anthus sylvanus* R Bulbul-: 17 cm. Steep grassy slopes, abandoned terrace cultivation and open pine forest. From the Sulaiman Range north to Swat and lower Kagan Valley, and east along the Himalayas to Nepal and Sikkim; between *c.* 900 and 3000 m.

13. (1886) Ssp. *Motacilla alba personata*]
14. (1885) Ssp. *Motacilla alba dukhunensis*] PIED or WHITE WAGTAIL *Motacilla alba* RM Bulbul-: 18 cm.
Open country in the vicinity of watercourses, reservoirs, rice and cultivation, lawns and playing fields. Subcontinent; Andamans. Vagrant or absent in Sri Lanka.

15. (1874) FOREST WAGTAIL *Motacilla indica* RM Sparrow+: 17 cm. Evergreen or moist-deciduous biotope: forest, coffee plantations, mixed bamboo jungle, etc. *Breeding*: N Cachar district, Assam. *Winter*: Nepal, Sikkim, NE India; Bangladesh; E and W Ghats complexes; Sri Lanka.

16. (1876) YELLOW WAGTAIL *Motacilla flava* RM Sparrow±, with longer tail: 17 cm. Pastureland, wet paddy stubbles, etc. *Breeding*: Ladakh; N Kashmir(?). *Winter*: Subcontinent; also Andamans, Nicobars; Sri Lanka.

17. (1878) Blackheaded ssp. *melanogrisea* of 1876. Subcontinent (winter).

18. (1891) LARGE PIED WAGTAIL *Motacilla maderaspatensis* R Bulbul±: 21 cm. Watercourses, rocky streams, reservoirs and pools. Continental and peninsular India.

19. (1884) GREY WAGTAIL *Motacilla cinerea* M Sparrow±, with longer tail: 17 cm. Along clear mountain streams with rocky banks and boulders. *Summer*: From N Baluchistan to Chitral, east through Baltistan, Gilgit, Ladakh and Kashmir to Nepal; between *c.* 1800 and 4300 m. *Winter*: Himalayan foothills, south through the Subcontinent; Andamans, Nicobars; Sri Lanka; Maldives.

20. (1883) YELLOWHEADED WAGTAIL *Motacilla citreola* RM Sparrow±, with longer tail: 17 cm. Marshes, irrigated ricefields, etc. *Summer*: N Baluchistan; NWFP, east through Baltistan, Gilgit, Kashmir, Ladakh, Lahul, Spiti; between *c.* 1500 and 4600 m. *Winter*: Subcontinent. Absent in Andaman and Nicobar Is.

PLATE 96

1. (986) GLOSSY STARE or STARLING *Aplonis panayensis* R? Myna±: 22 cm. Forest and coconut groves. Andaman and Nicobar Is.; Meghalaya and Assam south through Bangladesh (status uncertain).

2. (984) SPOTTEDWINGED STARE *Saroglossa spiloptera* M? Bulbul±: 19 cm. Open tall forest. *Summer*: Himalayas. *Winter*: Assam south to Bangladesh (Dhaka, Chittagong).

3. (987) GREYHEADED MYNA *Sturnus malabaricus* R Myna-: 21 cm. Thinly wooded country and secondary jungle. Continental and peninsular India. Has wandered in winter to Lower Sind (Thatta dist.) and Sri Lanka.

4. (988) Whiteheaded ssp. *blythi* of 987. SW India from Belgaum south.

5. (994) BLACKHEADED, or BRAHMINY, MYNA *Sturnus pagodarum* R Myna-: 2 cm. Open deciduous forest and scrub jungle, and near cultivation and habitations. Subcontinent; Sri Lanka.

6. (995) DAURIAN MYNA *Sturnus sturninus* V Myna-: 19 cm. Isolated specimens from Chitral, Andaman (on board ship), and Nicobar Is. (Camorta).

7. (993) CEYLON WHITEHEADED MYNA or STARLING *Sturnus senex* R (endemic) Myna-: 22 cm. Tall forest edges and clearings. SW Sri Lanka.

8. (991) WHITEHEADED MYNA *Sturnus erythropygius* R Myna-: 21 cm. Forest and secondary jungle. Andaman and Nicobar Is.

9. (997) STARLING *Sturnus vulgaris* M Bulbul±: 19-20 cm. Cultivations, orchards, marshes and vicinity of habitations. *Summer*: NWFP east through Kashmir. *Winter*: Pakistan and N India east to Bangladesh, south to Gujarat, Madhya Pradesh and Andhra Pradesh.

10. (996) ROSY PASTOR *Sturnus roseus* M Myna±: 23 cm. Open cultivation, grassland and semi-desert. Pakistan; continental and peninsular India; Sri Lanka; Bangladesh.

11. (1002) PIED MYNA *Sturnus contra* R Myna±: 23 cm. Open country and cultivation. From Haryana east to NE India and Bangladesh; Himalayan foothills, south through Madhya Pradesh to Deccan. Extending range into Pakistan.

12. (1006) COMMON MYNA *Acridotheres tristis* R Pigeon-: 23 cm. Neighbourhood of homesteads, villages and cities. Subcontinent; Sri Lanka. Introduced in the Andaman, Nicobar, Maldive and Lakshadweep Is.

13. (1005) CHINESE or GREYBACKED MYNA *Sturnus sinensis* V Bulbul±: 20 cm. Manipur (once).

14. (1008) BANK MYNA *Acridotheres ginginianus* R Myna-: 21 cm. Neighbourhood of human habitations. Pakistan; continental and peninsular India south to Maharashtra, Madhya Pradesh and Orissa.

15. (1009) JUNGLE MYNA *Acridotheres fuscus* R Myna±: 23 cm. Well-wooded country in the vicinity of cultivation and villages. Himalayas; NE hill states; Bangladesh (Chittagong). Patchily in continental and peninsular India.

16. (1014) GOLDCRESTED MYNA *Mino coronatus* R Myna±: 21 cm. Moist deciduous and semi-evergreen open forest and cultivation clearings. Manipur and Assam.

17. (1013) COLLARED MYNA *Acridotheres albocinctus* R Myna±: 23 cm. Meadows, elephant grass and vicinity of cultivation and villages. Manipur valley.

18. (1012) ORANGEBILLED JUNGLE MYNA *Acridotheres javanicus* R Myna±: 23 cm. Semi-cultivation, tall grass and near villages. Nagaland south through Manipur and Mizoram to the Chittagong hill tracts; up to *c.* 1300 m.

19. (1015) HILL MYNA *Gracula religiosa* R Myna±: 25-29 cm. Evergreen and semi-evergreen forest and vicinity of cultivation. Himalayan foothills from Kumaon east; NE hill states; Bangladesh (Chittagong); E and W Ghat complexes south of *c.* 17°N; Sri Lanka; Andaman and Nicobar Is.

20. (1019) CEYLON HILL MYNA *Gracula ptilogenys* R (endemic) Myna+: 25 cm. High trees in forest and cultivation. Sri Lanka.

PLATE 96
1
imm
2 ♂
♀
4
4
5 imm
6 imm
3
5
6
10 imm
10
7
9 imm
9 W
10
8
9 S
13 imm
11
12
13
14
15
12
16
17
18
20
19
John H. Dick

PLATE 97
1
2
3 ♂
♀
4 ♂
♀
5 ♂ S
5 ♂ W
♀
6 ♂
♀
7 ♂
♀
8 ♂
♀
9 ♂
♀
10 ♂
♀
11 ♂
♀
12 ♂
♀
13 ♂
♀
14
15
♀
16 ♂
John H. Dick

PLATE 97

1. (1933) WHITE-EYE *Zosterops palpebrosa* R Sparrow-: 10 cm. Forest, groves, gardens, orchards and mangrove. Subcontinent excepting desert parts; Andamans and Nicobar Is. Sri Lanka.

2. (1937) CEYLON WHITE-EYE *Zosterops ceylonensis* R (endemic) Sparrow-: 11 cm. Forest, tea plantations, gardens, etc. Sri Lanka.

3. (1906) RUBYCHEEK *Anthreptes singalensis* R Sparrow-: 10 cm. Dense evergreen forest, open scrub jungle. E Himalayas; NE hill states; Bangladesh. E Ghats (sight). Foothills to *c.* 700 m.

4. (1907, 1908) PURPLERUMPED SUNBIRD *Nectarinia zeylonica* R Sparrow-: 10 cm. Scrub, light secondary jungle, dry cultivated country, gardens. S Bengal; Bangladesh; peninsular India; Sri Lanka.

5. (1913 to 1915) OLIVEBACKED SUNBIRD *Nectarinia jugularis* R Sparrow-: 10 cm. Forest, scrub and mangroves. Andaman and Nicobar Is.

6. (1910) VAN HASSELT'S SUNBIRD *Nectarinia sperata* R Sparrow-: 9 cm. Forest, gardens, and dense cover in swampy land. Assam; NE hill states; Bangladesh.

7. (1909) SMALL SUNBIRD *Nectarinia minima* R Sparrow-: 8 cm. Forest, sholas, gardens, flowering trees and bushes in hilly evergreen biotope. W Ghats complex from about Nasik south; normally between 300 and 2000 m.

8. (1919) MRS GOULD'S SUNBIRD *Aethopyga gouldiae* R Sparrow-: Male 15 cm. Female 10 cm. Evergreen and moist-deciduous biotope: forest and scrub jungle. Himalayas; NE hill states; Bangladesh (Chittagong region). Seasonally, foothills to *c.* 3600 m.

9. (1911, 1912) LOTEN'S SUNBIRD *Nectarinia lotenia* R Sparrow-: 13 cm. Moist-deciduous biotope: well-wooded, open country, gardens and cultivation. Southern peninsular India; Sri Lanka.

10. (1917) PURPLE SUNBIRD *Nectarinia asiatica* R Sparrow-: 10 cm. Light deciduous or dry thorn forest, cultivation, gardens and compounds. Subcontinent; Sri Lanka. Absent in Andaman and Nicobar Is.

11. (1923) NEPAL YELLOWBACKED SUNBIRD *Aethopyga nipalensis* R Sparrow-: Male 15 cm; Female 10 cm. Heavy forest of oak and rhododendron, scrub jungle and gardens. Himalayas from Mussooree east; NE hill states; Bangladesh (Chittagong hill tracts). Seasonally, *c.* 300-3600 m.

12. (1925) BLACKBREASTED SUNBIRD *Aethopyga saturata* R Sparrow-: Male 15 cm; Female 10 cm. Evergreen and moist-deciduous biotope: dense scrub, secondary growth or open pastureland with scattered bushes. Himalayas from Mussooree; NE hill states; Bangladesh. Seasonally, foothills to *c.* 2000 m.

13. (1927 to 1929a) YELLOWBACKED SUNBIRD *Aethopyga siparaja* R Sparrow-; Male 15 cm; Female 10 cm. Dense evergreen and moist-deciduous open forest and scrub jungle, gardens, orchards, groves, etc. Himalayan foothills from Kangra east to Sikkim; Bangladesh; continental and NE India; W Ghats complex from the Narbada river south; Nicobar Is.

14. (1931) LITTLE SPIDERHUNTER *Arachnothera longirostris* R Sparrow-: 14 cm. Moist-deciduous and evergreen forest, secondary growth, heavy jungle and sholas. E Himalayan foothills; NE hill states; Bangladesh. Patchily also in W and E Ghat complexes; Orissa. Up to *c.* 1800 m.

15. (1932) STREAKED SPIDERHUNTER *Arachnothera magna* R Sparrow+: 17 cm. Evergreen biotope: dense forest and abandoned cultivation clearings overgrown by wild banana stands. E Himalayas; NE hill states; Bangladesh; up to *c.* 2200 m.

16. (1930) FIRETAILED SUNBIRD *Aethopyga ignicauda* R Sparrow-: Male 15 cm; Female 10 cm. Open coniferous forest with dense growth of dwarf rhododendron and juniper, etc. Himalayas from Garhwal east; NE hill states; Bangladesh. Seasonally, *c.* 1200-4000 m.

PLATE 98

1. (1892) THICKBILLED FLOWERPECKER *Dicaeum agile* R Sparrow-: 9 cm. Dry to moist-deciduous or semi-evergreen biotope: on flowering or fruiting trees and shrubs in forest and cultivated country. Subcontinent; Sri Lanka.

2. (1896) YELLOWBELLIED FLOWERPECKER *Dicaeum melanoxanthum* R Sparrow-: 12 cm. Tall trees in open forest and forest clearings. Himalayas from Kumaon east; NE hill states. Seasonally, *c.* 1050-3600 m.

3. (1899) TICKELL'S FLOWERPECKER *Dicaeum erythrorhynchos* R Sparrow-: 8 cm. Moist-deciduous biotope: forest plantations, groves, orchards and scrub jungle. Subcontinent excepting desert parts; Sri Lanka.

4. (1902) PLAINCOLOURED FLOWERPECKER *Dicaeum concolor* R Sparrow-: 8 cm. Deciduous and mixed deciduous-evergreen forest and groves. E Himalayas; NE hill states; Bangladesh; up to *c.* 1800 m; W Ghats complex from Mahabaleshwar south; Andaman Is.

5. (1895) YELLOWVENTED FLOWERPECKER *Dicaeum chrysorrheum* R Sparrow-: 9 cm. Open jungle, forest edges and orange orchards. E Himalayas; NE hill states; Bangladesh (Chittagong hill tracts). Foothills to *c.* 2000 m.

6. (1897) LEGGE'S FLOWERPECKER *Dicaeum vincens* R (endemic) Sparrow-: 9 cm. Tall trees and climbers in rain forest. Sri Lanka.

7. (1898) ORANGEBELLIED FLOWERPECKER *Dicaeum trigonostigma* R Sparrow-: 9 cm. Glades and margins of evergreen and tidal forest. Submontane tracts of Arunachal Pradesh and Assam; Sunderbans.

8. (1905) FIREBREASTED FLOWERPECKER *Dicaeum ignipectus* R Sparrow-: 7 cm. Tall forest, secondary growth and orchards, etc. Himalayas; NE hill states. Seasonally *c.* 600-2700 m.

9. (1964) RED MUNIA, or AVADAVAT *Estrilda amandava* R Sparrow-: 10 cm. Sub-Himalayan terai and duars; swampy reed-beds and tall grassland, sugarcane fields, gardens and around villages. Subcontinent. Not in Andamans, Nicobars and Sri Lanka.

10. (1904) SCARLETBACKED FLOWERPECKER *Dicaeum cruentatum* R Sparrow-: 7 cm. Open forest groves and orchards. E Himalayas, south through Assam, Bangladesh and Bengal; up to *c.* 1400 m.

11. (1977) NE ssp. *Lonchura malacca atricapilla*]
12. (1978) Peninsular ssp. *Lonchura malacca malacca*] BLACKHEADED MUNIA *Lonchura malacca* R Sparrow-: 10 cm. Cultivation, tall grass and swampy ground. N, NE and peninsular India; Bangladesh; Sri Lanka.

13. (1965) GREEN MUNIA *Estrilda formosa* R Sparrow-: 10 cm. Grass and low bushes, sugarcane fields and boulder-strewn scrub jungle. Mainly C India.

14. (1974) SPOTTED MUNIA *Lonchura punctulata* R Sparrow-: 10 cm. Open woodland, bush-clad hillsides, grassland, gardens and cultivation. Subcontinent, in Pakistan occupying a narrow foothill zone; Sri Lanka.

15. (1971) Ssp. *Lonchura kellarti jerdoni* ssp. of 1973. India.

16. (1973) RUFOUSBELLIED MUNIA *Lonchura kellarti* R Sparrow-: 10 cm. Scrub, grassland, fallow fields in forest clearings, cultivation and near settlements. Southern parts of E and W Ghat complexes; Sri Lanka.

17. (1966) WHITETHROATED MUNIA *Lonchura malabarica* R Sparrow-: 10 cm. Cultivation, grassland, sparsely scrubbed country and light secondary jungle. Foothills and plains, Pakistan; continental and peninsular India; Bangladesh; Sri Lanka.

18. (1978a.) JAVA SPARROW *Padda oryzivora* R (introduced). Sparrow±: 15 cm. Paddy fields, gardens and reed-beds. Colombo; Calcutta; Madras.

19. (1968) WHITEBACKED MUNIA *Lochura striata* R Sparrow-: 10 cm. Open country, light scrub jungle and secondary growth. Himalayan foothills from Garhwal east; continental and peninsular India; Sri Lanka; Andaman and Nicobar Is.

1
♀
2♂
3
4
♀
6 ♂
♀
♀
5
7♂
8♂
♀
imm
11imm
♀
10 ♂
11
12
9 ♂
14
15
13
imm
19
imm
16
17
18
19
John H. Dick

PLATE 99
♀
♀
♀
1 ♂
2 ♂
3 ♂
♀
♀
♀
4 ♂
5 ♂
6 ♂
7
8
♀
♀
♀
9 ♂
♂ W
10 ♂ br +
11 ♂ br +
13,14 ♀
14 ♂ br +
♂ W and ♀
12 ♂ br +
13 ♂ br +
John H. Dick

PLATE 99

1. (EL) SAXAUL SPARROW *Passer ammodendri* C Asia, Mongolia.

2. (1938) HOUSE SPARROW *Passer domesticus* R Bulbul-: 15 cm. A ubiquitous commensal of man in cities, suburbs, villages, etc. Subcontinent; Sri Lanka. Absent in Nicobar Is; Andamans (introduced).

3. (1940) SPANISH SPARROW *Passer hispaniolensis* M Bulbul-: 15 cm. Cultivation and semi-desert. Plains of NW Pakistan and NW India.

4. (1947a) SCRUB SPARROW *Passer moabiticus* M Sparrow-: 12 cm. Scrub of *Prosopsis, Rubus, Tamarix*, etc. N Baluchistan.

5. (1945) SIND JUNGLE SPARROW *Passer pyrrhonotus* R Sparrow-: 12 cm. Mixed tamarisk, acacia and grass jungle along rivers and canals, and around jheels and marshes. Flood plains of the Indus in Punjab and Sind. Also NW India.

6. (1946) CINNAMON TREE SPARROW *Passer rutilans* R Bulbul-: 15 cm. Light forest of oak, rhododendron, etc., and terraced cultivation near villages. Himalayas; NE hill states. Seasonally, foothills to *c.* 2700 m.

7. (1942) TREE SPARROW *Passer montanus* R Bulbul-: 15 cm. Villages and fields, near human settlement. N Baluchistan; Himalayas; NE hill states; E Ghats; Bangladesh. Foothills to *c.* 2700 m.

8. (1950) ROCK SPARROW *Petronia petronia* M Sparrow+: 17 cm. Stony or rocky ground. Gilgit, NWFP, south to Mianwali and Quetta.

9. (1949) YELLOWTHROATED SPARROW *Petronia xanthocollis* R Sparrow-: 14 cm. Dry-deciduous forest, scrub, thorn jungle, groves, cultivation, and oases with date palms. Subcontinent; Sri Lanka (rare).

10. (1961) BLACKTHROATED WEAVER BIRD *Ploceus benghalensis* R Sparrow±: 15 cm. Swampy reed-beds in cultivated lowlands. Chiefly N and NE and continental India; Bangladesh.

11. (1960) FINN'S BAYA *Ploceus megarhynchus* R Sparrow+: 16 cm. Sub-Himalayan terai country with marshes, sarpat and *Saccharum* grass sparsely dotted with isolated trees. Kumaon terai (below Naini Tal) east to Bengal and Assam.

12. (1962) STREAKED WEAVER BIRD *Ploceus manyar* R Sparrow±: 15 cm. Swampy reed-beds in cultivated lowland. Pakistan; N and NE India; Bangladesh; capriciously in the Peninsula; Sri Lanka.

13. (1957) BAYA *Ploceus philippinus* R Sparrow±: 15 cm. Open cultivation, grassland and secondary scrub dotted with acacias and palm trees, etc. Subcontinent; Sri Lanka.

14. (1959) Palethroated eastern ssp. *burmanicus* of 1957. E Nepal terai, E Bihar, Bengal, Bhutan and Bangladesh.

PLATE 100

1. (1777 to 1779) ALPINE ACCENTOR *Prunella collaris* R Sparrow+: 17 cm. Stony slopes, cliffs and moraines. Himalayas, seasonally *c.* 1800-5500 m.

2. (1783) RUFOUSBREASTED ACCENTOR *Prunella strophiata* R Sparrow±: 15 cm. Conifer, oak or birch forest and scrub; boulder-strewn alpine meadows. Himalayas, seasonally *c.* 460-5000 m.

3. (1781) ROBIN ACCENTOR *Prunella rubeculoides* R Sparrow+: 17 cm. Dwarf willows and furze patches near streams, tundra-like vegetation around lakes. Tibetan facies of Himalayas from Baltistan, Astor and Ladakh. Seasonally, *c.* 2500-5300 m.

4. (1788) MAROONBACKED ACCENTOR *Prunella immaculata* R Sparrow±: 15 cm. Humid, mossy, conifer and rhododendron forest; secondary jungle, forest margins, etc. (winter).Himalayas from Garhwal east, seasonally *c.* 2100-4200 m.

5. (1780) ALTAI ACCENTOR *Prunella himalayana* M Sparrow±: 15 cm. Bare rocky hillsides. Himalayas, *c.* 2000-4200 m.

6. (1787) BLACKTHROATED ACCENTOR *Prunella atrogularis* M Sparrow±: 15 cm. Scrub jungle on hillsides, tea gardens, orchards, and bushes near cultivation. N Baluchistan, NW Pakistan and east along Himalayan foothills to NW Nepal.

7. (1785a) Ssp. *ocularis* of 1784. V Chaman, Baluchistan (once), Mashelakh plain, N Quetta.

8. (1787a) SIBERIAN ACCENTOR *Prunella montanella* V Sparrow±: 15 cm. Bushes and forest edges, near streams. Ladakh (once).

9. (1784) BROWN ACCENTOR *Prunella fulvescens* R Sparrow+: 15 cm. *Caragana* furze bushes on mountain slopes, other bushes in dry facies. *Summer*: Hunza, Astor, Baltistan and Ladakh to Rupshu; *c.* 3300-5100 m. *Winter*: Lahul, Chitral and Quetta, Nepal and Sikkim; Arunachal (?).

10. (1955) BLANFORD'S SNOW FINCH *Montifringilla blanfordi* M Sparrow±: 15 cm. Tibetan steppe country and cultivation near upland villages. N Sikkim; presumably also Ladakh.

11. (1956) PERE DAVID'S SNOW FINCH *Montifringilla davidiana* V Sparrow+: 15 cm. Open desert in high Tibetan facies. N Sikkim (once).

12. (1952) TIBET SNOW FINCH *Montifringilla adamsi* R Sparrow+: 17 cm. Stony plateaux, boulder-strewn hillsides and neighbourhood of upland villages. *Breeding*: Ladakh, Spiti, Nepal and N Sikkim; *c.* 3600-4900 m. *Winter*: Kulu, upper Sutlej valley.

13. (1951) PALLAS'S SNOW FINCH *Montifringilla nivalis* Status? Sparrow+: 17 cm. Occurs in Afghanistan, only in the Hindu Kush range. May extend to the adjacent Safed Koh.

14. (1953) MANDELLI'S SNOW FINCH *Montifringilla taczanowskii* M Sparrow+: 17 cm. Tibetan steppe facies. N Sikkim; possibly also Ladakh.

15. (2000) HODGSON'S MOUNTAIN FINCH *Leucosticte nemoricola* R Sparrow±: 15 cm. Alpine meadows, screes, moraines and dwarf scrub. Also open forested slopes and fallow fields (winter). Himalayas, seasonally *c.* 1000-5300 m.

16. (1954) REDNECKED SNOW FINCH *Montifringilla ruficollis* M Sparrow±: 15 cm. Open gravel plains, grassy plateaux and banks of meandering streams. N Sikkim; Darjeeling (once).

17. (2003) BRANDT'S MOUNTAIN FINCH *Leucosticte brandti* RM Sparrow+: 18 cm. Desolate stony hillsides, scree fans, moraines and alpine meadows. From Chitral east along Karakoram and northern Himalayas through Baltistan, Ladakh, Lahul, Spiti, Nepal and Sikkim. Seasonally *c.* 1500-5400 m.

18. (EL) BARTAILED SNOW FINCH *Montifringilla theresae* Afghanistan, Transcaspia.

1 2 3 7 6 5 4 8 9 10 11 13 12 10 13 12 14 15 imm 14 16 W 17 S W 18 S

John H. Dick

PLATE 101
♀
♀
imm
♂ W
3
1 ♂
♀
2 ♂ S
♀
4 ♂
6
imm
5 ♂
6 ♂
7
10
10
♀
8
9
8 ♂
♀
12 ♂ S
♀
11 ♂
♀
14 ♂
♀
13 ♂ S
♀
15 ♂
John H. Dick

PLATE 101

1. (1979) CHAFFINCH *Fringilla coelebs* M Sparrow±: 15 cm. Orchards. Kohat, Gilgit, Uttar Pradesh, NW Nepal.

2. (1980) BRAMBLING *Fringilla montifringilla* M Sparrow±: 15 cm. Gardens and orchards. N Baluchistan, NWFP, Gilgit and Kashmir, straggling to Simla and Mussooree; NW Nepal.

3. (1998) GOLDFRONTED FINCH *Serinus pusillus* R sparrow-: 12 cm. Rocky hillsides with stunted bushes, near cultivation, shingle screes and dwarf junipers. N Baluchistan; Chitral and east along the Himalayas to Nepal. Seasonally, *c.* 750-4700 m.

4. (1993) TIBETAN SISKIN *Serinus thibetanus* M Sparrow-: 12 cm. Alders, hemlock, birch and mixed fir forest. E Himalayas, seasonally *c.* 1000-3800 m.

5. (EL) EUROPEAN GREENFINCH *Carduelis chloris* SW Asia, Afghanistan.

6. (1990) HIMALAYAN GREENFINCH *Carduelis spinoides* RM Sparrow-: 14 cm. Open slopes, forest edges, open conifer forest and scrub, cultivation. Himalayas; Nagaland; Manipur. Seasonally, foothills to *c.* 4400 m.

7. (EL) BLACKHEADED GREENFINCH *Carduelis ambigua* SE Tibet, N Burma.

8. (1994) LINNET *Acanthis cannabina* M Sparrow-: 13 cm. Open country and stony slopes up to *c.* 2400 m. N Baluchistan; Gilgit; Baltistan; Kashmir; NW Nepal.

9. (1995, 1996) TWITE *Acanthis flavirostris* R Sparrow-: 13 cm. Stony and grassy slopes in arid country and boulder-strewn alpine meadows. Chitral; Gilgit; Baltistan; Ladakh; Spiti; Nepal; N Sikkim. Seasonally, *c.*1500-4800 m.

10. (1989) GOLDFINCH *Carduelis carduelis* R Sparrow-: 14 cm. Orchards, bare stony hillsides, open conifer forest, fields and scrub. N Baluchistan; W Himalayas. Seasonally foothills to *c.* 3900 m.

11. (2008) LICHTENSTEIN'S DESERT FINCH *Rhodospiza obsoleta* R Sparrow±: 15 cm. Orchards, fallow fields, weed-patches, etc. C Baluchistan to Chitral; up to *c* . 1400 m.

12. (2009) CRIMSONWINGED DESERT FINCH *Callacanthis sanguinea* Status? Bulbul-: 18 cm. Semi-desert, rocky and scrub-covered hillsides. Chitral and Ladakh (once each).

13. (2006) TRUMPETER BULLFINCH *Carpodacus githagineus* R Sparrow±: 15 cm. Bare hills and stony semi-desert. Baluchistan, Sind and NWFP north to Chitral. *Winter*: Makran coast, Rajasthan, Salt Range and Punjab.

14. (2007) MONGOLIAN TRUMPETER BULLFINCH *Carpodacus mongolicus* M Sparrow±: 15 cm. Dry rocky or stony slopes and steep ravines. Chitral, Gilgit, Astor, Baltistan, Ladakh, Baluchistan (Quetta); *c.* 1500-3000 m.

15. (EL) SINAI ROSEFINCH *Carpodacus synoicus* NE Afghanistan.

PLATE 102

1. (2016) BLANFORD'S ROSEFINCH *Carpodacus rubescens* R Sparrow±: 15 cm. Conifer or mixed conifer and birch forest. E Himalayas, seasonally *c.* 1300-3900 m.

2. (2015) NEPAL ROSEFINCH *Carpodacus nipalensis* R Sparrow±: 15 cm. Forest of oak, rhododendron and silver fir; grassy slopes with stunted bushes and stony pastures. Himalayas, seasonally *c.* 1200-4400 m.

3. (2013) COMMON ROSEFINCH, SCARLET GROSBEAK *Carpodacus erythrinus* RM Sparrow±: 15 cm. Willows along streams and river-beds, rocky bush-covered slopes, thorny scrub, etc. *Summer*: N Baluchistan; W Himalayas. *Winter*: Pakistan south to Sind; Himalayan foothills; continental and peninsular India; Bangladesh.

4. (2017) PINKBROWED ROSEFINCH *Carpodacus rhodochrous* R Sparrow±: 15 cm. Open fir and birch forest, willow bushes and dwarf juniper; also grassy slopes, gardens and upland villages. Himalayas from Indus Kohistan east, seasonally *c.* 600-4200 m.

5. (2023) BEAUTIFUL ROSEFINCH *Carpodacus pulcherrimus* R Sparrow±: 15 cm. Rhododendron and other bushes on steep hillsides in dry biotope. Himalayas from Garhwal east; seasonally *c.* 1800-4500 m.

6. (2025) LARGE ROSEFINCH *Carpodacus edwardsii* R Sparrow±: 17 cm. Rhododendron and silver fir forest; open forest and mountains with scrub (winter). E Himalayas, seasonally *c.* 1000-3900 m.

7. (2017a) VINACEOUS ROSEFINCH *Carpodacus vinaceus* Status? Sparrow-: 13 cm. Dense bushes and clearings in bamboo forest. Himalayas from Naini Tal (Kumaon) to Nepal, where possibly breeding.

8. (2019) SPOTTEDWINGED ROSEFINCH *Carpodacus rhodopeplus* R Sparrow±: 15 cm. Grassy hillsides with bushes. Himalayas from E Kumaon through Nepal and Sikkim.

9. (2026) THREEBANDED ROSEFINCH *Carpodacus trifasciatus* M Bulbul-: 17 cm. Orchards and hedges in cultivated fields. SE Tibet along the Tsangpo; presumably also Arunachal Pradesh.

10. (2021) WHITEBROWED ROSEFINCH *Carpodacus thura* R Sparrow+: 17 cm. Light fir, juniper and rhododendron forest, and dwarf juniper or rhododendron. Himalayas, from NWFP east; seasonally *c.* 1800-4200 m.

11. (2020) *Carpodacus thura blythi* ssp. of 2021. Himalayas from NWFP east to Garhwal.

12. (2028) EASTERN GREAT ROSEFINCH *Carpodacus rubicilloides* R Bulbul±: 19 cm. *Caragana*, willow and *Hippophae* scrub in arid country. Northern Himalayas and Tibetan plateau from Ladakh east to Bhutan. Breeding *c.* 3700-5200 m.

13. (2018) REDMANTLED ROSEFINCH *Carpodacus rhodochlamys* R Bulbul-: 18 cm. Bushes and shrubs in dry biotope; thorny scrub, gardens and cultivation (winter). N Baluchistan; W Himalayas. Seasonally, foothills to *c.* 3800 m.

14. (2033) REDHEADED ROSEFINCH *Propyrrhula subhimachala* R Bulbul±: 20 cm. Dense rhododendron, juniper and willow scrub and light forest with dense bush undergrowth. Himalayas from Nandadevi to the east; NE hill states. Seasonally, *c.* 1800-4200 m.

15. (2027) GREAT ROSEFINCH *Carpodacus rubicilla* R Bulbul±: 19 cm. Boulder-strewn ground with sparse vegetation at the foot of mountains. From Chitral east through Hunza, Gilgit, Ladakh, Lahul, Spiti, Nepal and Sikkim. Seasonally *c.* 1500-5000 m.

16. (2031) REDBREASTED ROSEFINCH *Carpodacus puniceus* R Bulbul±: 20 cm. Steep, rugged mountainsides. Himalayas from Gilgit east through Baltistan, Ladakh and Kashmir to Arunachal Pradesh. Seasonally, *c.* 1500-5200 m.

17. (2032) CROSSBILL *Loxia curvirostra* R Sparrow±: 15 cm. Coniferous forest. From Himachal Pradesh (Lahul and Kulu) east through Bhutan and Arunachal Pradesh; *c.* 2700-4000 m.

PLATE 102
♀
♀
♀
1 ♂
2 ♂
3 ♂
♀
4 ♂
♀
♀
♀
5 ♂
6 ♂
7 ♂
♀
8 ♂
9 ♂
10 ♀
11 ♀
♀
10 ♂
♀
♀
12 ♂
13 ♂
14 ♂
♀
♀
♀
15 ♂
16 ♂
17 ♂

imm
1
imm
♀
2 ♂
imm
♀
3 ♂
imm
♀
imm
4 ♂
imm
imm
5 ♂
6 ♂
imm
♀
♀
♀
♀
9
7 ♂
8 ♂
imm
10 ♂
♀
♀
♀
11 ♂
12 ♂
13 ♂
imm
♀W
♀S
♀
♀
14 ♂
15 ♂
16 ♂

PLATE 103

1. (2041) CORN BUNTING *Emberiza calandra* V Bulbul-: 18 cm. Waste ground. Jhang district, Pakistan (once).

2. (EL) YELLOWHAMMER *Emberiza citrinella* Europe, W Asia.

3. (2042) PINE BUNTING *Emberiza leucocephalos* M Sparrow+: 17 cm. Bush-covered grassy slopes and cultivation. Baluchistan north through NWFP and east through Salt Range, Gilgit, Kashmir and the Himalayan foothills to Nepal; up to *c.* 1500 m.

4. (2050) GREYNECKED BUNTING *Emberiza buchnani* M Sparrow±: 15 cm. Dry stony foothills with sparse bushes; sparsely shrubbed country and broken hillsides (winter). *Summer*: N Baluchistan and the Afghan Safed Koh. *Winter*: W, N and C India south to Karnataka. Sparsely in Sind, but numerous on autumn and spring passage.

5. (2048) WHITECAPPED BUNTING *Emberiza stewarti* M Sparrow±: 15 cm. Rocky or grassy hillsides, scrub jungle. *Summer*: N Baluchistan; W Himalayas. *Winter*: NWFP and Salt Range east to Nepal; the plains of NW India south to Gujarat, NE Maharashtra; Kirthar Range in Sind. Seasonally, foothills to *c.* 3600 m.

6. (2051 to 2054) ROCK BUNTING *Emberiza cia* RM Sparrow±: 15 cm. Grassy and bush-clad slopes in boulder-strewn country, juniper and open pine forest. W Himalayas; SE Tibet and Arunachal Pradesh. *Winter*: Pakistan and N India. Seasonally, foothills to *c.* 4200 m.

7. (2049) ORTOLAN BUNTING *Emberiza hortulana* V Sparrow±: 15 cm. Orchards or grassy slopes with bushes. Recorded only from Gilgit (twice), Kashmir (once) and Delhi (once).

8. (2057) STRIOLATED BUNTING *Emberiza striolata* R Sparrow-: 14 cm. Rocky or stony, sparsely scrubbed hillsides and nullahs, ancient forts and ruins, etc. S Baluchistan, Sind and Punjab Salt Range; N and C India, and C Maharashtra.

9. (2056) LITTLE BUNTING *Emberiza pusilla* M Sparrow-: 14 cm. Reeds, grass, rice stubbles and scrub, about cultivation in open country. W Himalayas (rare); E Himalayas; NE hill states; Bangladesh (hill tracts).

10. (2055) GREYHEADED BUNTING *Emberiza fucata* RM Sparrow±: 15 cm. Bushes on hillsides, along rivers and on swampy ground. W Himalayas, seasonally, foothills to *c.* 2700 m. *Winter*: NE India, Bangladesh; Nepal.

11. (2046) YELLOWBREASTED BUNTING *Emberiza aureola* M Sparrow±: 15 cm. Cultivation and grassland, hedgerows, gardens, etc. Nepal, Sikkim, NE India, Bangladesh, N Bengal; up to *c.* 1500 m. A record each from Baluchistan and Nicobars.

12. (2043) BLACKHEADED BUNTING *Emberiza melanocephala* M Sparrow+: 18 cm. Cereal cultivation. Pakistan; N, W and C Indian plains; Nepal (once); also Gilgit, Himachal and Punjab.

13. (2045) CHESTNUT BUNTING *Emberiza rutila* M Sparrow±: 14 cm. Rice stubble, bushes in cultivation and forest clearings. Nepal, Sikkim, Jalpaiguri duars, Manipur, N Cachar; Chitral (once).

14. (2047) BLACKFACED BUNTING *Emberiza spodocephala* M Sparrow±: 15 cm. High grass, bamboo and scrub jungle, hedgerows, ricefields and reedy marshes, up to *c.* 1000 m. NE India and Bangladesh; also the Jalpaiguri and Sikkim duars, west to Kathmandu valley.

15. (2058) REED BUNTING *Emberiza schoeniclus* M Sparrow-: 14 cm. Reed-beds, tall grass and bush jungle in riverain country. Sind, N Baluchistan, NWFP, Punjab, Gilgit and NW India in Punjab and Haryana.

16. (2044) REDHEADED BUNTING *Emberiza bruniceps* RM Sparrow+: 17 cm. Cultivation, especially grain fields. *Summer*: N Baluchistan. *Winter*: Plains of N, W and C India; Bangladesh.

PLATE 104

1. (794) HONEYGUIDE *Indicator xanthonotus* R Sparrow±: 15 cm. Conifer, dry deciduous or subtropical wet forest with cliffs and combs of rocks bees. Himalayas from Hazara east; Nagaland; Manipur; *c.* 1500-3500 m.

2. (2034) SCARLET FINCH *Haematospiza sipahi* R Bulbul-: 18 cm. Open conifer forest. Himalayas from Garhwal east through Bhutan and Arunachal Pradesh, south through Meghalaya; *c.* 1600-2400 m.

3. (1997) REDBROWED FINCH *Callacanthis burtoni* R Sparrow+: 17 cm. Open forest of fir, pine, deodar or birch. Himalayas from Safed Koh east to Sikkim, *c.* 800-3000 m.

4. (2036) BROWN BULLFINCH *Pyrrhula nipalensis* R Sparrow+: 17 cm. Dense fir, oak and rhododendron forest. Himalayas; NE hill states; *c.* 1500-3900 m.

5. (2040) ORANGE BULLFINCH *Pyrrhula aurantiaca* R Sparrow±: 14 cm. Open fir, birch and mixed forest. W Himalayas, *c.* 1600-2700 m.

6. (2039) REDHEADED BULLFINCH *Pyrrhula erythrocephala* R Sparrow+: 17 cm. Deciduous forest. Himalayas, *c.* 1500-4200 m.

7. (2038) BEAVAN'S BULLFINCH *Pyrrhula erythaca* R Sparrow+: 17 cm. Conifer and rhododendron forest. E Himalayas, *c.* 2000-3800 m.

8. (1981) HAWFINCH *Coccothraustes coccothraustes* M Sparrow±: 18 cm. Groves, orchards, gardens and wooded foothills. N Baluchistan; NWFP.

9. (1982) BLACK-AND-YELLOW GROSBEAK *Coccothraustes icterioides* R Myna±: 22 cm. Pine, silver fir and deodar forest. W Himalayas, *c.* 1800-3500 m.

10. (1983) ALLIED GROSBEAK *Coccothraustes affinis* R Myna±: 22 cm, Oak, rhododendron, or mixed conifer and broadleafed forest. Himalayas, *c.* 1800-4200.

11. (1986) SPOTTEDWINGED GROSBEAK *Coccothraustes melanozanthos* R Myna±: 22 cm. Mixed conifer and broadleafed forest. Himalayas, *c.* 600-3600 m.

12. (2035) GOLDHEADED BLACK FINCH *Pyrrhoplectes epauletta* R Sparrow±: 15 cm. Rhododendron and ringal bamboo undergrowth. Himalayas from Simla east; *c.* 1400-3600 m.

13. (1985) WHITEWINGED GROSBEAK *Coccothraustes carnipes* R Myna±: 22 cm. Fir, rhododendron or juniper and mixed forest. Baluchistan; Himalayas; *c.* 1500-4200 m.

14. (2060) CRESTED BUNTING *Melophus lathami* R Sparrow±: 15 cm. Dry, stony, sparsely scrubbed hillsides; rice stubbles and open stony dry-deciduous jungle and charred grassland (winter). Himalayan foothills from Hazara; NE hill states; Bangladesh; most of continental and peninsular India. Up to *c.* 1800 m.

1
♀
♀
4
2 ♂
3 ♂
♀
♀
♀
♂ imm
5 ♂
6 ♂
7 ♂
8
♀
♀
9 ♂
10 ♂
♀
♀
11 ♂
12 ♂
♀
13 ♂
14 ♂
♀
John H. Dick

PLATE 105

WILLIAMSI
1
LEUCOMELANA
2
MELANOTA
3
LATHAMI
4
MOFFITTI
5
J. H. Dick

PLATE 105

Subspecies of the Kaleej Pheasant (*Lophura leucomelana*)

1. (297) WILLIAMS'S KALEEJ *L.l.williamsi* R Size same as 294. Habitat as in other subspecies. Possibly extends from contiguous Chin Hills area of Burma into E Manipur and Mizo.

2. (294) NEPAL KALEEJ *L.l.leucomelana* R Village hen±: 60-68 cm. See Plate 35.

3. (295) BLACKBACKED KALEEJ *L.l.melanota* R Size same as 294. Overgrown ravines on hill slopes. Himalayas 600-2700 m; E Nepal to W Bhutan.

4. (296) BLACKBREASTED KALEEJ *L.l.lathami* R Size same as 294. Overgrown ravines and gullies on hill slopes. Himalayas 100-2600 m; E Bhutan eastward; NE hill states; Bangladesh (Chittagong).

5. (298) MOFFITT'S KALEEJ *L.l.moffitti* R Size same as 294. Habitat unknown: presumably same as in other subspecies. Distribution enigmatic as type specimen came from animal dealer. Now proved to be an intergrade between 295 and 296 (qq.v).

PLATE 106

Tails of Snipes (Genus *Gallinago*)
(Only half shaded)

1. (408) GREAT SNIPE *Gallinago media* V Quail+: 27 cm. Marshland. S India; Sri Lanka; Andamans (winter).

2. (407) SWINHOE'S SNIPE *Gallinago megala* M Quail+: 29 cm. Marshy edges of jheels, wet paddy stubbles, etc. Subcontinent; Sri Lanka; Maldives (winter).

3. (406) PINTAIL SNIPE *Gallinago stenura*
For distribution see Plate 41

4. (405) WOOD SNIPE *Gallinago nemoricola* RM Partridge±: 31 cm. Swampy patches amidst tall grass and scrub in hilly country. Himalayas from Kulu east, 2100-4200 m. *Winter*: Lower Himalayas; NE hill states; peninsular India; Sri Lanka.

5. (409) FANTAIL SNIPE *Gallinago gallinago*
For distribution see Plates 41, 43

6. (404) SOLITARY SNIPE *Gallinago solitaria* RM Partridge±: 31 cm. Sprawling boggy mountain streams interspersed with grassy hummocks. Himalayas 2400-4600 m; down to base of hills and plains level (winter).

7. (410) JACK SNIPE *Gallinago minima*
For distribution see Plate 43

Tails of Snipes Genus (*Gallinago*)

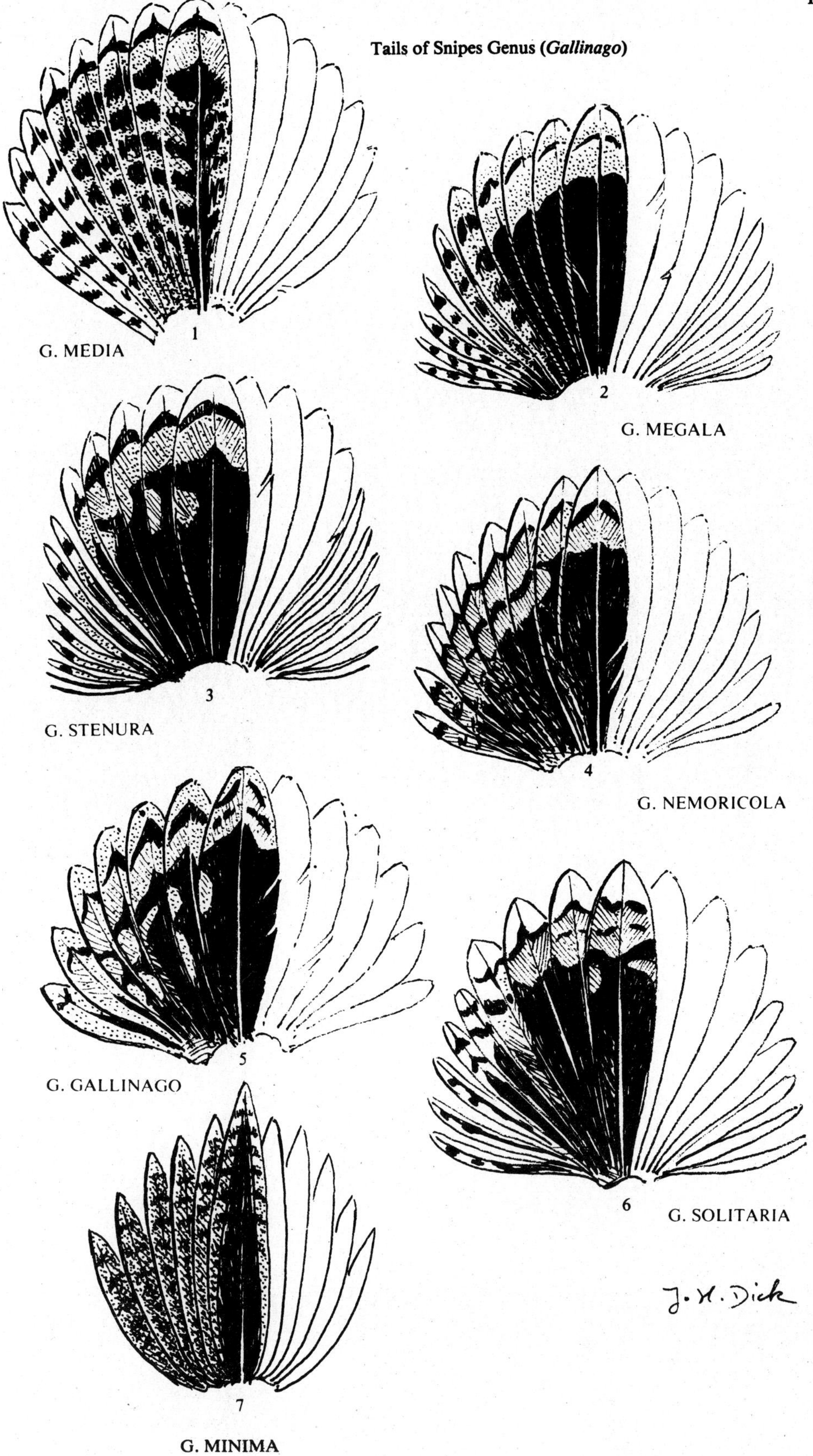

ALPHABETICAL INDEX OF THE FAMILIES AND SPECIES GROUPS

NOTES

NOTES